T0122621

Intelligent Systems Reference Library

Volume 86

Series editors

Janusz Kacprzyk, Polish Academy of Sciences, Warsaw, Poland
e-mail: kacprzyk@ibspan.waw.pl

Lakhmi C. Jain, University of Canberra, Canberra, Australia, and
University of South Australia, Adelaide, Australia
e-mail: Lakhmi.Jain@unisa.edu.au

About this Series

The aim of this series is to publish a Reference Library, including novel advances and developments in all aspects of Intelligent Systems in an easily accessible and well structured form. The series includes reference works, handbooks, compendia, textbooks, well-structured monographs, dictionaries, and encyclopedias. It contains well integrated knowledge and current information in the field of Intelligent Systems. The series covers the theory, applications, and design methods of Intelligent Systems. Virtually all disciplines such as engineering, computer science, avionics, business, e-commerce, environment, healthcare, physics and life science are included.

More information about this series at http://www.springer.com/series/8578

Anand Jayant Kulkarni
Kang Tai · Ajith Abraham

Probability Collectives

A Distributed Multi-agent System Approach for Optimization

Anand Jayant Kulkarni
School of Mechanical and Aerospace Engineering
Nanyang Technological University
Singapore
Singapore

Kang Tai
School of Mechanical and Aerospace Engineering
Nanyang Technological University
Singapore
Singapore

Ajith Abraham
Scientific Network for Innovation and Research Excellence
Machine Intelligence Research Labs (MIR Labs)
Auburn, WA
USA

ISSN 1868-4394 ISSN 1868-4408 (electronic)
Intelligent Systems Reference Library
ISBN 978-3-319-36521-3 ISBN 978-3-319-16000-9 (eBook)
DOI 10.1007/978-3-319-16000-9

Springer Cham Heidelberg New York Dordrecht London
© Springer International Publishing Switzerland 2015
Softcover reprint of the hardcover 1st edition 2015
This work is subject to copyright. All rights are reserved by the Publisher, whether the whole or part of the material is concerned, specifically the rights of translation, reprinting, reuse of illustrations, recitation, broadcasting, reproduction on microfilms or in any other physical way, and transmission or information storage and retrieval, electronic adaptation, computer software, or by similar or dissimilar methodology now known or hereafter developed.
The use of general descriptive names, registered names, trademarks, service marks, etc. in this publication does not imply, even in the absence of a specific statement, that such names are exempt from the relevant protective laws and regulations and therefore free for general use.
The publisher, the authors and the editors are safe to assume that the advice and information in this book are believed to be true and accurate at the date of publication. Neither the publisher nor the authors or the editors give a warranty, express or implied, with respect to the material contained herein or for any errors or omissions that may have been made.

Printed on acid-free paper

Springer International Publishing AG Switzerland is part of Springer Science+Business Media (www.springer.com)

Preface

This book is written for engineers, scientists, and students studying/working in the Optimization, Artificial Intelligence (AI) or Computational Intelligence (CI) arena, and particularly involved in the Collective Intelligence (COIN) field. Mainly, the book in detail provides the core and underlying principles and analysis of the different concepts associated with an emerging AI tool in the framework of COIN for modeling and controlling distributed Multi-Agent Systems (MAS) referred to as Probability Collectives (PC). The tool was first proposed by David Wolpert in 1999 in a technical report presented to NASA and was further elaborated by Stefan Bieniawski in 2005.

More specifically, the book in detail discusses the modified PC approach proposed by the authors of this book. The modifications reduced the computational complexity and improved the convergence and efficiency of the PC methodology. In order to further extend the PC approach and make it more generic and powerful, a number of constraint handling techniques are incorporated into the overall framework to develop the capability for solving constrained problems since real-world practical problems are inevitably constrained problems. In the course of these modifications, various inherent characteristics of the PC methodology are thoroughly explored, investigated, and validated. The book demonstrates the validation of the modified PC approach by successfully optimizing several unconstrained test problems. The first constrained PC approach exploits various problem-specific heuristics for successfully solving two test cases of the Multi-Depot Multiple Traveling Salesmen Problem (MDMTSP) and several cases of the Single Depot MTSP (SDMTSP). The second constrained PC approach incorporating penalty functions into the PC framework is tested by solving a number of constrained test problems. In addition, two variations of the feasibility-based rule for handling constraints are proposed and are tested solving two cases of the Circle Packing Problem (CPP) as well as various cases of the Sensor Network Coverage Problem (SNCP). The results highlighted the robustness of the PC algorithm solving all the cases of the SNCP. In addition, PC with feasibility-based rule is successfully applied for solving several discrete and mixed variable problems in the structural and mechanical engineering domain.

The mathematical level in all the chapters is well within the grasp of the scientists as well as the undergraduate and graduate students from the engineering and computer science streams. The reader is encouraged to have basic knowledge of probability and mathematical analysis. In presenting the PC and associated modifications and contributions, emphasis is placed on development of the fundamental results from basic concepts. Numerous examples/problems are worked out in the text to illustrate the discussion. These illustrative examples may allow the reader to gain further insight into the associated concepts.

Over a period of 7 years, the material has been tested extensively and published in various prestigious journals and conferences. The suggestions and criticism of various reviewers and colleagues had a significant influence on the way the work has been presented in this book.

We are much grateful to our friends and colleagues for reviewing the different parts of the manuscript and for providing us valuable feedback. The authors would like to thank Dr. Thomas Ditzinger, Springer Engineering In-house Editor, Studies in Computational Intelligence Series, Professor Janusz Kacprzyk (Editor-in-Chief, Springer Studies in Computational Intelligence Series), and Mr. Holger Schäpe (Editorial Assistant, Springer Verlag, Heidelberg) for the editorial assistance and excellent cooperative collaboration to produce this important scientific work. We hope that the reader will share our excitement to present this volume on 'Metaheuristic Clustering' and will find it useful.

Singapore, November 2014 Anand Jayant Kulkarni
Singapore Kang Tai
Auburn Ajith Abraham

Contents

Chapter 1
Introduction to Optimization

1.1 What is Optimization?

For almost all human activities and creations, there is a desire to do or be the best in some sense. To set a record in a race, for example, the aim is to do the fastest (shortest time); in the conduct of the retail business, the desire may be to maximize profits; in the construction of a building, the desire may be to minimize costs; in the planning of a project schedule, the aim may be to minimize project time; in the design of a power generator turbine, the objective may be to maximize efficiency; and in the design of VLSI integrated circuits, the aim may be to lay out as many transistors as possible within each IC chip.

Hence the concept of minimization and maximization has great significance in both human affairs and the laws of nature. Optimization therefore refers to a positive and intrinsically human concept of minimization or maximization to achieve the best or most favorable outcome from a given situation. In addition, as the element of design is present in all fields of human activity, all aspects of optimization can be viewed and studied as design optimization without any loss of generality. This makes it clear that the study of design optimization can help not only in the human activity of creating optimum design of products, processes and systems, but also in the understanding and analysis of mathematical/physical phenomenon and in the solution of mathematical problems.

There are always numerous requirements and constraints imposed on the design of components, products, processes or systems in real-life engineering practice, just as in all other fields of design activity. Therefore, creating a feasible design under all these diverse requirements/constraints is already a difficult task, and to ensure that the feasible design created is also 'the best' is even more difficult.

© Springer International Publishing Switzerland 2015

A.J. Kulkarni et al., *Probability Collectives*, Intelligent Systems Reference Library 86, DOI 10.1007/978-3-319-16000-9_1

1.1.1 General Problem Statement

All the optimal design problems can be expressed in a standard general form stated as follows:

$$\text{Minimize objective function} \quad G(\mathbf{X}) \tag{1.1}$$

Subject to

$$s \text{ number of inequality constraints } g_j(\mathbf{X}) \le 0, \quad j = 1, 2, \ldots, s \tag{1.2}$$

$$w \text{ number of equality constraints } h_j(\mathbf{X}) = 0, \quad j = 1, 2, \ldots, w \tag{1.3}$$

where the number of design variables is given by $x_i, \quad i = 1, 2, \ldots, n$

or by design variable vector $\mathbf{X} = \begin{Bmatrix} x_1 \\ x_2 \\ \vdots \\ x_n \end{Bmatrix}$

- A problem where the objective function is to be maximized (instead of minimized) can also be handled with this standard problem statement since maximization of a function $G(\mathbf{X})$ is the same as minimizing the negative of $G(\mathbf{X})$.
- Similarly, the '$\ge$' type of inequality constraints can be treated by reversing the sign of the constraint function to form the '$\le$' type of inequality.
- Sometimes there may be simple limits on the allowable range of value a design variable can take, and these are known as side constraints:

$$x_i^l \le x_i \le x_i^u$$

  where x_i^l and x_i^u are the lower and upper limits of x_i, respectively. However, these side constraints can be easily converted into the normal inequality constraints (by splitting them into 2 inequality constraints).
- Although all optimal design problems can be expressed in the above standard form, some categories of problems may be expressed in alternative specialized forms for greater convenience and efficiency.

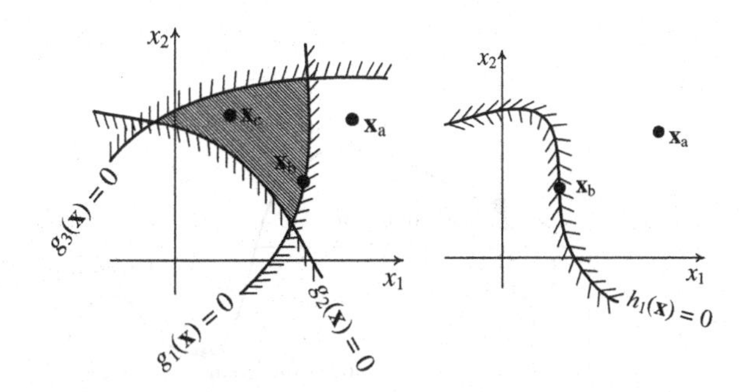

Fig. 1.1 Active/inactive/violated constraints

1.1.2 Active/Inactive/Violated Constraints

The constraints in an optimal design problem restrict the entire design space into smaller subset known as the feasible region, i.e. not every point in the design space is feasible. See Fig. 1.1.

- An inequality constraint $g_j(\mathbf{X})$ is said to be violated at the point $\mathbf{x}$ if it is not satisfied there $(g_j(\mathbf{X}) \geq 0)$.
- If $g_j(\mathbf{X})$ is strictly satisfied $(g_j(\mathbf{X}) < 0)$ then it is said to be inactive at $\mathbf{x}$.
- If $g_j(\mathbf{X})$ is satisfied at equality $(g_j(\mathbf{X}) = 0)$ then it is said to be active at $\mathbf{x}$.
- The set of points at which an inequality constraint is active forms a constraint boundary which separates the feasibility region of points from the infeasible region.
- Based on the above definitions, equality constraints can only be either violated $(h_j(\mathbf{X}) \neq 0)$ or active $(h_j(\mathbf{X}) = 0)$ at any point $\mathbf{x}$.
- The set of points where an equality constraint is active forms a sort of boundary both sides of which are infeasible.

1.1.3 Global and Local Minimum Points

Let the set of design variables that give rise to a minimum of the objective function $G(\mathbf{X})$ be denoted by $\mathbf{X}^*$ (the asterisk $*$ is used to indicate quantities and terms referring to an optimum point). An objective $G(\mathbf{X})$ is at its global (or absolute) minimum at the point $\mathbf{X}^*$ if:

$$G(\mathbf{X}^*) \leq G(\mathbf{X}) \quad \text{for all } \mathbf{X} \text{ in the feasible region}$$

The objective has a local (or relative) minimum at the point $\mathbf{X}^*$ if:

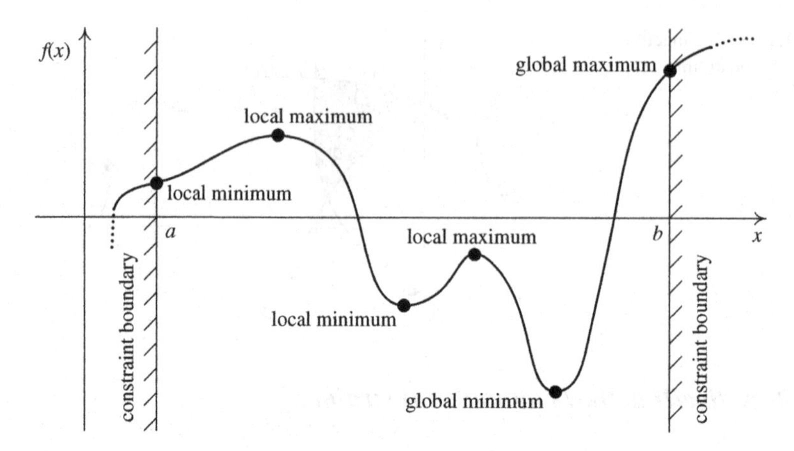

Fig. 1.2 Minimum and maximum points

$$G(\mathbf{X}^*) \leq G(\mathbf{X}) \quad \text{for all feasible } \mathbf{X}$$
$$\text{within a small neighborhood of } \mathbf{X}^*$$

A graphical representation of these concepts is shown in Fig. 1.2 for the case of a single variable x over a closed feasible region $a \leq x \leq b$.

1.2 Contemporary Optimization Approaches

In past few years a number of nature-/bio-inspired optimization techniques such as Evolutionary Algorithms (EAs), Swarm Intelligence (SI), etc. have been developed. The EA such as Genetic Algorithm (GA) works on the principle of Darwinian theory of survival of the fittest individual in the population. The population is evolved using the operators such as selection, crossover, mutation, etc. According to Deb [1] and Ray et al. [2], GA can often reach very close to the global optimal solution and necessitates local improvement techniques to incorporate into it. Similar to GA, mutation driven approach of Differential Evolution (DE) was proposed by Storn and Price [3] which helps explore and further locally exploit the solution space to reach the global optimum. Although, easy to implement, there are several problem dependent parameters required to be tuned and may also require several associated trials to be performed.

Inspired from social behavior of living organisms such as insects, fishes, etc. which can communicate with one another either directly or indirectly the paradigm of SI is a decentralized self organizing optimization approach. These algorithms work on the cooperating behavior of the organisms rather than competition amongst them. In SI, every individual evolves itself by sharing the information from others in the society. The techniques such as Particle Swarm Optimization (PSO) is inspired from the social behavior of bird flocking and school of fish searching for food [4]. The Ant Colony Optimization (ACO) works on the ants' social behavior of foraging food following a shortest path [5]. Similar to ACO, the Bee Algorithm (BA) also works on the social behavior of honey bees finding the food; however, the bee colony tends to optimize the use of number of members involved in particular predecided tasks [6]. The Firefly Algorithm (FA) is an emerging metaheuristic SI technique based on the idealized behavior of the flashing characteristics of fireflies [7, 8]. Generally, the swarm techniques are computationally intensive.

1.3 Complex Systems and Significance of Distributed Optimization

A Complex System is a broad term encompassing a research approach to problems in the diverse disciplines such as neurosciences, social sciences, meteorology, chemistry, physics, computer science, psychology, artificial life, evolutionary computation, economics, earthquake prediction, molecular biology, etc. Generally it includes many components that not only interact but also compete with one another to deliver the best they can to reach the desired system objective. Moreover, it is difficult to understand the whole system only by knowing each component and its individual behavior. This is because any move by a component affects the further decisions/moves by other components and so on. There are many complex systems in engineering such as Internet Search, Engineering Design, Manufacturing and Scheduling, Logistics, Sensor Networks, Vehicle Routing, Aerospace Systems, etc.

Traditionally, such complex systems were seen as centralized systems, but as the complexity grew, it became tedious to be handled using the above discussed algorithms, i.e. in a centralized way and hence a distributed and decentralized optimization approach became necessary. In a distributed and decentralized approach, the system is divided into smaller subsystems and optimized individually to get the system level optimum. Such subsystems together can be seen as a collective which is a group of self-interested learning agents. Such a group can be referred to as a Multi-Agent System (MAS). In the Collective Intelligence (COIN) framework, these systems can be optimized individually to attain the system level optimum. In a distributed MAS, the rational and self-interested behavior of the agents is very important to achieve the best possible local goal/reward/payoff, but it is not trivial to make such agents work collectively to achieve the best possible global or system objective. The significance of a distributed, decentralized and cooperative approach which underscores its superiority over the centralized approach is discussed below.

In a centralized system as shown in Fig. 1.3, a single agent is supposed to have all the capabilities, such as problem solving, in order to alleviate the user's cognitive load. The agent is provided with the general knowledge which is useful to do a wide variety of tasks.

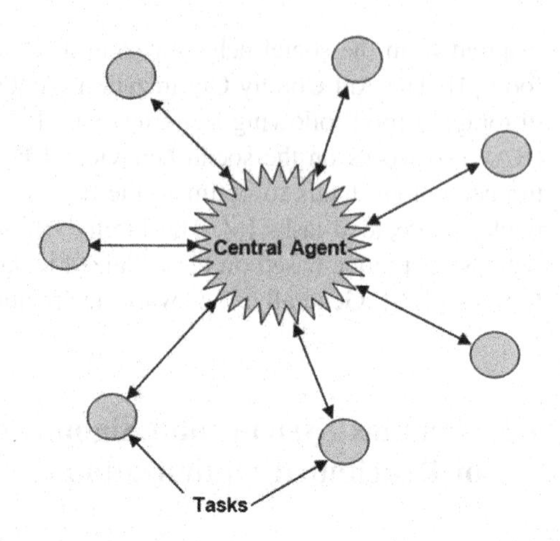

Fig. 1.3 A centralized system

This single agent needs enormous knowledge to deal effectively with the user requests; it needs to do required computations and also needs enough storage space, etc. Such a system may have a processing bottleneck. It also cannot be a robust system because of the possibility of single point failure which may have serious impact. For example, the process of information search over internet, information finding, filtering, etc. may overwhelm a centralized system. On the other hand, if the work is divided or decomposed into different tasks (e.g. information search, filtering, evaluation and integration) by classifying the expertise needed in the system, the potential bottleneck can be avoided. The classification of expertise here refers to the decomposition of the system into various agents representing experts of the particular area in the system. This expertise is the set of knowledge or actions the agent is supposed to have under the particular circumstance. It is therefore natural to have a distributed MAS having expertise for different heterogeneous information sources [9]. Working ants and social bees are examples of natural biological MASs. In these biological MASs the bees and ants work independently with a global goal/system objective such as the collection of food.

Furthermore, it is evident that such distributed, decentralized MAS is intended for open-ended design problems where there is no well-defined solution available and requires an in-depth search of the design space and also requires adaptability to user preferences. Designing a MAS to perform well on a collective platform is non-trivial. Moreover, the straightforward agent learning in a MAS on collective platform cannot be implemented as it can lead to a suboptimal solution and also may interfere with individual interests. This is referred to as 'Tragedy of Commons' in which the rational and self-interested independent individuals deplete the shared limited resource in a greedy way, even if it is well understood that it may not be beneficial for long term interest collectively for all, i.e. an individual may receive a benefit but on the other hand the loss will be shared among all. This conflict may further lead to total system collapse. It also highlights the inadequacy of the classic Reinforcement Learning (RL) approach. An attractive option is to devise a distributed system in which different parts are referred to as agents; each having local control, having cooperation among one another and contributing towards a common aim. The cooperation among such agents is also important in order to avoid the replication of the work and/or information and reduction in the computational load.

In the case of such cooperative distributed and decentralized system, there is also a need for the right to share the information among the agents. It is important when dealing with security/military systems, banking systems, construction systems, etc. When these systems are used as centralized systems, the information from all the individual processes need to be communicated to the centralized controller, planner, scheduler, etc. This can be a hindrance when there may be some subsystems not willing to share the information with the third party (or a centralized system). Such systems are very difficult to be optimized centrally unless the right to share the information or right to cooperate is clearly defined. The centralized approach is not suitable for the modern highly dynamic environment as by the time the re-computation and re-distribution is done the environment is already changed. This makes it clear that a centralized system may add latency in the overall processing. On the other hand, the sub-problems can naturally be distributed and decentralized into agents and on the basis of the inter-agent-cooperation, quicker decisions can be taken and executed. The autonomous agents perform computations independently from other agents, and contribute their results in a parallel and distributed fashion. Such collaborating agents make the system robust and flexible to adapt easily to the user preference changes. It is also believed that the cooperating agents in a distributed system collectively show better performance than a centralized system or a distributed system with agents having absolutely no cooperation.

Furthermore, the major challenges addressed in the collective behavior of the autonomous agents are how to enable the distributed agents to dynamically acquire their goal-directed cooperative behavior in performing a certain task, and how to apply the collective learning when addressing an ill-defined problem. The crucial step is to formulate and develop the structure and organize the agents in a MAS ensuring achievement of the global optimum using the local knowledge. The other challenges are how to ensure the robustness, how to impart adaptability, how to maintain privacy, etc.

1.3.1 Advantages of the Distributed, Decentralized and Cooperative Approach

The above approach is used in variegated applications because of its various advantages. Some of the important advantages are discussed here.

1. It reduces the dependence on a single central system, thus reducing the chance of single point failure. This imparts the important characteristic referred to as

robustness. This is essential in the field of UAV as failure of the centralized controller can be devastating and may result in collision.
2. It reduces computational overhead on a centralized system.
3. It reduces communication delays between resource agents and the coordinator as the information is transferred to the corresponding agents rather than the central system/agent. It also solves the security issue such as sharing of information with a centralized system as the exchange of local information is allowed between the entitled sub-systems/agents.
4. It allows the planning to be done on a shorter time scale, allowing additional planning/coordination rounds to take place.
5. It reduces complexity of the coordination problem making the controllers simpler and cheaper.
6. It reduces overall complexity of the problem formulation.
7. It imparts modularity and flexibility as one can reconfigure or expand the system by adding new components, sub-systems in the realm of the problem and still the computations can be carried out locally and no single system overload occurs. This feature is referred to as scalability.
8. As the computations are done at the local level, the system can be made more flexible by dividing it into different problem solvers based on the abilities. This way the decision making becomes faster by exploiting parallelism.
9. It enhances the local communication and subsequently reduces the overwhelming energy consumption. A sensor network communicating the raw data to the central point uses more energy as compared to local communication in the distributed approach.
10. The large and complex systems can be built into individual and simpler modules which are easier to debug and maintain.
11. It imparts reliability through redundancy.
12. It expedites the decision-making process by reducing latency.
13. It helps to improve the real time response and behavior.

Some of the important applications exploiting the above advantages are discussed below.

1.4 Prominent Applications of the Distributed, Decentralized and Cooperative Approach

The complex adaptive systems which work on distributed, decentralized and cooperative MASs, because of their various advantages, have been successfully applied to various engineering applications such as internet search, engineering design, manufacturing, scheduling, logistics, sensor networks, vehicle routing, UAV path planning and many more are currently being researched. Some of the applications are discussed below.

Meta-heuristics is the field in which several search processes interchange the information while searching the optimal solution. To have a successful search process, the information exchange is very important. In case of a centralized system there is a central agent that carries out the interchange of the information between the various processes. In the decentralized cooperative system, each process has its own rules to decide when, what and how to interchange the relevant information with the relevant processes [9]. This approach of local communication reduces computational overload for the system, thus speeding up further decision making. According to Batista et al. [10], logistics problems such as the Vehicle Routing Problem (VRP), loading problem (Bin Packing) and location problem can be solved using a decentralized cooperative way.

The information sources available online are inherently distributed and are having different modalities. It is therefore natural to have a distributed MAS having expertise in different heterogeneous information sources. The robustness of such a system is most important because online information services are dynamic and unstable in nature. Sycara et al. [9] presented a multi-agent computational infrastructure referred to as Reusable Task Structure-based Intelligent Network Agents (Retsina) which searches information on the internet, filters out irrelevant information, integrates information from the heterogeneous sources and also updates the stored information in a distributed manner.

It is true that a sensor network communicating the raw data to the central point uses more energy as compared to the distributed approach [11]. This forces undesirable limits on the amount of information to be collected. This also imposes constraints on the amount of information to be communicated as it affects communication overload. In the case of distributed and decentralized sensor networks, each node is an individual solving the problems independently and cooperatively. Similarly, Petcu [12] proposed a distributed and decentralized Radio Frequency Identification (RFID) system in which a subsystem or a sensor is not only a node collecting and transferring the data but also does the processing and computing job in order to take decisions locally. These subsystems, nodes or sensors are equipped with as much knowledge, logic, and rights as possible to make it a better alternative to the centralized system.

Campbell et al. [13] presented a MAS the agents in which could handle the pre- and post-processing of various computational analysis tools such as spreadsheets or CAD systems in order to have a common communication between them. These agents communicate through a common framework where they act as experts and communicate their results to the centralized design process. As the number of design components increase, the number of agents and the complexity also increases. This results in the growing need for communication, making it computationally cumbersome for a centralized system. This is one of the reasons that the centralized approach is becoming insignificant in the concurrent design.

The construction industries are becoming more complex and involving many subcontractors. In a project, subcontractors perform most of the work independently and manage their own resources. Their work affects the other subcontractors and eventually the entire project. This makes it clear that the centralized system or control of such process is inadequate to handle such situations. This suggests that the control needs to be distributed into individual sub-contractors in order to reschedule their projects dynamically [14].

Kim and Paulson [14] illustrated a distinguished MAS referred to as Distributed Coordination Framework for Project Schedule Changes (DCPSC). The objective is the minimization of the total extra cost each sub-contractor has to incur because of the abrupt project schedule change. Every sub-contractor is assumed to be a software agent to enhance communication using the internet. The agents in this distributed system compete and take socially rational decisions to maintain a logical sequence of the network. As the project needs to be rescheduled dynamically, extensive communication/negotiation among the subcontractor agents is required. In this approach these agents interact with one another to evaluate the impact of changes, simulate decisions in order to reduce the possible individual loss.

There is a growing trend in many countries to generate and distribute power/energy locally using renewable and non-conventional sources [15]. They are 'distributed' because they are placed at or near the point of energy consumption, unlike traditional 'centralized' systems where electricity/energy is generated at a remotely located, large-scale power plant and then transmitted down power lines to the consumer [16]. The same is true for the distribution of heating and cooling systems [15]. Depending on the peak and non-peak periods, the supply of energy has to be changed dynamically to accommodate the unforeseen behavior and current demand characteristics.

The distributed and decentralized systems related to robotic systems need very high robustness and flexibility. Examples of such applications are semi-automatic space exploration [17], rescue [18], and underwater exploration [19]. The most common technique to ensure the robustness in robotic systems is to decompose the complex system into a distributed and decentralized system and also to introduce redundancy like that of wireless networks [11, 20, 21] in which even though some nodes fail, the network continues to operate [22]. Based on the experimentation, it is claimed that robotic hardware redundancy is necessary along with distributed and decentralized control [23]. It is worth to highlight that the self-reconfigurable robots such as MTRN [24] and PolyBot [25] use centralized control despite a good hardware flexibility. As these robots are less robust to failures, this disadvantage overwhelms flexibility.

Unmanned Vehicles (UVs) or Unmanned Aerial Vehicles (UAVs) are seen as flexible and useful for many applications such as searching targets, mapping a given area, traffic surveillance, fire monitoring, etc. [26]. They are very useful in the environment where the use of manned airplane mission may be dangerous or impossible. UV/UAV is a strong potential field in which decentralized and distributed approach can be applied for conflict resolution, collision avoidance, better utilization of air-space and mapping efficient trajectory. Some of these applications are discussed below.

Krozel et al. [27] demonstrated the advantages of the distributed and decentralized approach when applied to Air/Ground Traffic Management. Based on the specialized job, the system is decomposed into three types of agents: airplanes/flight deck, air traffic service provider and airline operational control. In the traditional approach of Air/Ground Traffic Management the focus typically is on smooth, stable and orderly flow of traffic at the expense of efficiency. Moreover, the modern concept of free flight was also highlighted with emphasis on avoiding the latency in decision making in highly dense traffic conditions. If the planned trajectories are in conflict zone with one another the corresponding airplanes change the trajectories without communicating to the centralized system, avoiding communication overhead and further delay in the decision. This makes the centralized system insignificant keeping the communication to local level. Such decentralized control strategy reduces conflict alerts (crisis) as well.

In the dynamic environment of a war field, the risk or potential threats change their positions with respect to time. The objective of the UAVs is to reach a particular point at the same time following the shortest path. Zengin and Dogan [28] proposed a Probabilistic Map approach, in which based on the local information and/or positions of the threats the local updating of the terrain map was done and was directly communicated with the neighboring UAVs. This cooperation helped real time updating of the map and also decentralized and distributed control by each airplane was possible. Durfee et al. [29] proposed a similar approach in the sensor networks domain to find out the location of the airplane. The sensor nodes positioned topographically at different locations monitor the same aircraft but the perception of each node is different which may not necessarily provide information about the aircraft movement/location unless the information gathered by all these sensors is integrated and communicated to neighboring nodes to produce an overall picture of aircraft movements, i.e. the overall solution.

Anderson and Robbins [30] addressed the problem of of UAVs formation following a stationary/moving target. In the formation, each UAV tries to keep a desired distance from the neighboring one. It was claimed that forming the objective function controlling all the UAVs in the flock showing the formation instincts such as collision avoidance, obstacle avoidance, formation keeping, etc. is non-trivial. Also, dealing with such single objective function is computationally expensive. This becomes worse when the formation tries to add more UAVs. It is important to mention that each aircraft was modeled separately for collision avoidance, formation keeping, target seeking, obstacle avoidance, etc. Chang et al. [31] implemented a similar approach for a fleet of Multiple UAVs (MUAVs). In this model, gyroscopic force was used to change the direction of individual airplanes to avoid possible collisions. In both of these proposed works, as each vehicle has local control and computations, the system shows high scalability to include a large number of vehicles in the fleet. Similar to this, Garagic and Mehra [32] proposed computationally Distributed, Decentralized Optimization for Cooperative Control of Multi-agent Swarm-like Systems.

The number of potential conflicts grows exponentially as the number of vehicles grows. This highlights the insignificance of centralized approach and shows the need for a distributed and decentralized control approach. The space under consideration may get added with more vehicles, increasing the possibility of conflict. For handling this problem, Chanyagom et al. [33] developed a model of field sensing approach having attraction towards the target and repulsion to the obstacle or neighboring vehicle. Each UAV was modeled as a magnetic dipole and was provided with a sensor to detect the magnetic field generated by other UAVs estimating the gradient. This estimate was used to go in the opposite direction to avoid conflict. There was absolutely no cooperation among the UAVs. Wolpert et al. [34] successfully implemented a complex distributed system for solving an airplane fleet assignment problem. A detailed discussion of this approach is provided in the next chapter on Probability Collectives (PC).

The PC methodology in COIN framework through which the distributed, decentralized and cooperative optimization can be implemented is discussed in detail in the following chapter.

References

1. Deb, K.: An efficient constraint handling method for genetic algorithms. Comput. Methods Appl. Mech. Eng. **186**, 311–338 (2000)
2. Ray, T., Tai, K., Seow, K.C.: Multiobjective design optimization by an evolutionary algorithm. Eng. Optim. **33**(4), 399–424 (2001)
3. Storn, R., Price, K.: Differential evolution—a simple and efficient heuristic for global optimization over continuous spaces. J. Global Optim. **11**, 341–359 (1997)
4. Kennedy, J., Eberhart, R.: Particle swarm optimization. In: Proceedings of IEEE International Conference on Neural Networks, pp. 1942–1948 (1995)
5. Dorigo, M., Birattari, M., Stitzle, T.: Ant colony optimization: artificial ants as a computational intelligence technique. IEEE Computational Intelligence Magazine, pp. 28–39 (2006)
6. Pham, D.T., Ghanbarzadeh, A., Koc, E., Otri, S., Rahim, S., Zaidi, M.: The bees algorithm. Technical note, Manufacturing Engineering Centre, Cardiff University (2005)
7. Deshpande, A.M., Phatnani, G.M., Kulkarni, A.J.: Constraint handling in firefly algorithm. In: Proceedings of IEEE International Conference on Cybernetics, pp. 186–190 (2013)
8. Yang, X.S.: Firefly algorithms for multimodal optimization. In: Stochastic Algorithms: Foundations and Applications, SAGA, Lecture Notes in Computer Sciences, vol. 5792, pp. 169–178 (2009)
9. Sycara, K., Pannu, A., Williamson, M., Zeng, D., Decker, K.: Distributed intelligent agents. IEEE Expert Intell. Syst. Appl. **11**(6), 36–46 (1996)
10. Batista, B.M., Moreno Perez, J.A., Moreno Vega, J.M.: Nature-inspired decentralized cooperative metaheuristic strategies for logistic problems. In: Proceedings of European Symposium on Nature-Inspired Smart Information Systems (2006)
11. Rabbat, M., Nowak, R.: Distributed optimization in sensor networks. In: Proceedings of Third International Symposium on Information Processing in Sensor Networks, pp. 20–27 (2004)
12. Petcu, A.: Artificial intelligence laboratory. EPFL, Switzerland. http://liawww.epfl.ch/Research/dcr/. Accessed 08 July 2011
13. Campbell, M., Cagan, J., Kotovsky, K.: A-Design: an agent-based approach to conceptual design in a dynamic environment. Res. Eng. Des. **11**, 172–192 (1999)
14. Kim, K., Paulson, B.C.: Multi-agent distributed coordination of project schedule changes. Comput. Aided Civ. Infrastruct. Eng. **18**, 412–425 (2003)
15. http://www.det.csiro.au/science/de_s/index.htm. Accessed 08 July 2011
16. http://www.nrel.gov/learning/eds_distributed_energy.html. Accessed 08 July 2011

17. Visentin, G.,Winnendael, M.V., Putz, P.: Advanced mechatronics in ESA space robotics developments. In: Proceedings of the IEEE/RSJ International Conference on Intelligent Robots and Systems, vol. 2, pp. 1261–1266 (2001)
18. Casper, J., Murphy, R.R., Micire, M.: Issues in intelligent robots for search and rescue. Proc. SPIE Unmanned Ground Veh. Technol. **2**(4024), 292–302 (2000)
19. Ayers, J., Zavracky, P., McGruer, N., Massa, D.P., Vorus, W.S., Mukherjee, R., Currie, S.N.: A modular behavioral based architecture for biomimetic autonomous underwater robots. In: Proceedings of the Autonomous Vehicles in Mine Countermeasures Symposium, Naval Postgraduate School: Monterey, California, USA (1998)
20. Waldock, A., Nicholson, D.: Cooperative decision strategies applied to a distributed sensor network. In: Proceedings of 2nd SEAS DTC Technical Conference—Edinburgh (2007)
21. Lesser, V.R., Erman, L.D.: Distributed interpretation: a model and experiment. IEEE Trans. Comput. **C-29**(12):1144–1163 (1980)
22. Stengel, R.F.: Intelligent failure-tolerant control. IEEE Control Syst. Mag. **11**(4):14–23 (1991)
23. Mondada, F., Pettinaro, G.C., Guignard, A., Kwee, I.W., Floreano, D., Deneubourg, J., Nolfi, S., Gambardella, L.M., Dorigo, M.: Swarm-bot: a new distributed robotic concept. Auton. Robots 17:193–221 (2004)
24. Kamimura, A., Murata, S., Yoshida, E., Kurokawa, H., Tomita, K., Kokaji, S.: Self-reconfigurable modular robot experiments on reconfiguration and locomotion. In: Proceedings of the 2001 IEEE/RSJ International Conference on Intelligent Robots and Systems IROS2001, vol. 1, pp. 606–612 (2001)
25. Duff, D., Yim, M., Roufas, K.: Evolution of polybot: a modular reconfigurable robot. In: Proceedings of COE/Super Mechano-systems Workshop, Tokyo, Japan (2001)
26. Rathinam, S., Sengupta, R., Darbha, S.: A resource allocation algorithm for multivehicle systems with nonholonomic constraints. IEEE Trans. Autom. Sci. Eng. **4**(1), 98–104 (2007)
27. Krozel, J., Peters, M., Bilimoria, K.: A decentralized control strategy for distributed air/ground traffic separation. In: Proceedings of AIAA Guidance, Navigation and Control Conference and Exhibit, Paper, No. 4062 (2000)
28. Zengin, U., Dogan, A.: Probabilistic trajectory planning for UAVs in dynamic environments. In: Proceedings of AIAA 3rd Unmanned-Unlimited Technical Conference, Workshop, and Exhibit 2, Paper, No. 6528 (2004)
29. Durfee, E.H., Lesser, V.R., Corkill, D.D.: Trends in cooperative distributed problem solving. IEEE Trans. Knowl. Data Eng. **1**(1), 63–83 (1989)
30. Anderson, M.R., Robbins, A.C.: Formation flight as a cooperative game. In: Proceedings of AIAA Guidance, Navigation, and Control Conference, pp. 244–251 (1998)
31. Chang, D.E., Shadden, Marsden J.E., Olfati-Saber, R.: Collision avoidance for multiple agent systems. In: Proceedings of 42nd IEEE Conference on Decision and Control, pp. 539–543 (2003)
32. Garagic, D., Mehra, R.K.: Distributed, decentralized optimization for cooperative control of multi-agent swarm-like systems. Final report, NAVY Phase 1 STTR, Contract No. N00014-04-M-0289 (2005)
33. Chanyagom, P., Szu, H.H., Wang, H.: Collective behavior implementation in powerline surveillance sensor network. In: Proceedings of International Joint Conference on Neural Networks, pp. 1735–1739 (2005)
34. Wolpert, D.H., Antoine, N.E., Bieniawski, S.R., Kroo, I.R.: Fleet assignment using collective intelligence. In: Proceedings of the 42nd AIAA Aerospace Science Meeting Exhibit (2004)

Chapter 2
Probability Collectives: A Distributed Optimization Approach

An emerging Artificial Intelligence tool in the framework of Collective Intelligence (COIN) for modeling and controlling distributed Multi-agent System (MAS) referred to as Probability Collectives (PC) was first proposed by Dr. David Wolpert in 1999 in a technical report presented to NASA [1]. It is inspired from a sociophysics viewpoint with deep connections to Game Theory, Statistical Physics, and Optimization [2–4]. From another viewpoint, the method of PC theory is an efficient way of sampling the joint probability space, converting the problem into the convex space of probability distribution. PC considers the variables in the system as individual agents/players in a game being played iteratively [2, 5, 6]. Unlike stochastic approaches such as Genetic Algorithm (GA), Swarm Optimization or Simulated Annealing (SA), rather than deciding on the agent's moves/set of actions, PC allocates the probability values for selecting each of the agent's moves. At each iteration, every agent independently updates its own probability distribution over a strategy set which is the set of moves/actions affecting its local goal which in turn also affects the global or system objective [3]. The process continues and reaches equilibrium when no further increase in reward is possible for the individual agent by changing its actions further. This equilibrium concept is referred to as Nash equilibrium [5]. The concept is successfully formalized and implemented through the PC methodology. The approach works on probability distributions, directly incorporating uncertainty and is based on prior knowledge of the actions/strategies of all the other agents. The available literature on PC is discussed in the following section.

2.1 Background of PC

The approach of PC has been tested proving its variegated application domains and comparing its performance with other algorithms. The literature is discussed in the following few paragraphs.

Huang and Chang [7] demonstrated the search process of PC methodology is more robust as compared with the GA optimizing the multimodal function such as the Schaffer's function. In addition, it was also demonstrated that PC also

© Springer International Publishing Switzerland 2015

A.J. Kulkarni et al., *Probability Collectives*, Intelligent Systems Reference Library 86, DOI 10.1007/978-3-319-16000-9_2

outperformed GA in the rate of descent, trapping in false minima and long term optimization when tested and compared for the multimodality, nonlinearity and non-separability in solving several other benchmark problems such as Rosenbrock function, Ackley Path function and Michalewicz Epistatic function. Furthermore, Huang et al. [8] also discussed fundamental differences between GA and PC. At the core of the GA optimization algorithm is the population of solutions. In every iteration, each individual solution from the population is tested for its fitness to the problem at hand and the population is updated accordingly. GA plots the best-so-far curve showing the fitness of the best individual in the last preset generations. In PC, on the other hand, the probability distribution of the possible solutions is updated iteratively. After a predefined number of iterations, the probability distribution of the available strategies across the variable space is plotted in PC when optimizing an associated Homotopy function. It also directly incorporates uncertainty due to both imperfect sampling and the stochastic independence of the agents' actions. Furthermore, PC outperformed the Heuristic Adaptive GA (HAGA) [9], Heuristic ACO (HACO) [10] and Heuristic PSO (HPSO) [11] in stability and robustness solving the air combat decision-making for coordinated multiple target assignment problem [12]. The above comparison indicated that PC can potentially be applied to wide application areas.

Vasirani and Ossowski [13] solved the 8-queens problem and underscored the superiority of the decentralized PC architecture over a centralized one. They also demonstrated that both the approaches differ from each other because of the distributed sample generation and updating of the probabilities in the former approach. In addition, PC was also compared with the backtracking algorithm referred to as Asynchronous Distributed OPTimization (ADOPT) [14]. Although the ADOPT algorithm is a distributed approach, the communication and computational load was not equally distributed among the agents. It was also demonstrated that although ADOPT was guaranteed to find the solution in each run, communication and computations required were more than for the same problem solved using PC.

Wolpert et al. [5] successfully applied the PC methodology for solving the complex combinatorial optimization problem of airplane fleet assignment having the goal of minimization of the number of flights with 129 variables and 184 constraints. Applying a centralized approach to this problem may increase the communication and computational load. Furthermore, it may add latency in the system and result in the growing possibility of conflict in schedules and continuity. Using PC, the goal was collectively achieved by exploiting the advantages of a distributed and decentralized approach by the airplanes selecting their own schedules depending upon the individual payoffs for the possible routes. Mohammad and Babak [15] as well as Ryder and Ross [16] successfully applied the approach of PC solving combinatorial optimization problems such as the joint optimization of the routing and resource allocation in wireless networks.

Furthermore, Wolpert et al. [6] also tested the potential of PC in mechanical design domain for optimizing the cross-sections of individual bars and segments of a 10 bar truss. The problem was formulated as a discrete problem and the solution was feasible but was worse than those obtained by other methods [17–19]. In addition, Autry [20] also tested the PC methodology on the discrete problem of university course scheduling; however, the implementation failed to generate any feasible solution. It is important to note that although the problems in [5, 6, 20] are constrained problems, the constraints were not explicitly treated or incorporated in the PC algorithms.

Recently, Sislak et al. [21] proposed decentralized and semi-centralized approaches for solving an autonomous conflict resolution problem for cooperating airplanes to avoid the mid-air collision. In the first approach, every airplane was assumed to be an autonomous agent. These agents selected their individual paths and avoided collision with other airplanes traveling in the neighborhood. In order to implement this approach, a complex negotiation mechanism was required for the airplanes to communicate and cooperate with one another. In the second approach which is a semi-centralized approach, every airplane was given a chance to become a host airplane which computed and distributed the solution to all other airplanes. It is important to mention that the host airplane computed the solution based on the independent solution shared by previous host airplane. This process continued in a sequence until all the airplanes selected their individual paths. In both the approaches the airplanes were equidistantly arranged on the periphery of a circle. The targets of the individual airplanes were set as the opposite points on the periphery of the circle, setting the center point of the circle as a point of conflict and collision constraints were molded into special penalty functions which were further integrated into the objective function using suitable balancing factors. Similar to the weighted sum multi-objective approach [22, 23], the balancing factors were assigned based on the importance of the corresponding constraints. Smyrnakis and Leslie [24] proposed a Sequentially Updated PC (SPC) which was tested by optimizing the unconstrained Hartman's functions and the vehicle target assignment type of game. The SPC performed better with higher dimension Hartman's functions only but failed to converge in the target assignment game.

2.2 Conceptual Framework of PC

PC treats the variables in an optimization problem as individual self interested learning agents/players of a game being played iteratively. While working in some definite direction, these agents select actions over a particular interval and receive some local rewards on the basis of the system objective achieved because of those actions. In other words, these agents optimize their local rewards or payoffs, which also optimize the system level performance. The process iterates and reaches equilibrium also referred to as Nash equilibrium (See Sect. 2.2.2 for proof and details) when no further increase in the reward is possible for the individual agent through changing its actions further. Moreover, the method of PC theory is an efficient way of sampling the joint probability space, converting the problem into the convex space of probability distribution. PC allocates probability values to each agent's moves, and hence directly incorporates uncertainty. This is based on prior knowledge of the recent action or behavior selected by all other agents. In short, the agents in the PC framework need to have knowledge of the environment along with every other agent's recent action or behavior.

In every iteration, each agent randomly samples from within its own strategy set as well as from within other agents' strategy sets and computes the corresponding system objectives. The other agents' strategy sets are modeled by each agent based on their recent actions or behavior only, i.e. based on partial knowledge. By minimizing the collection of system objectives, every agent identifies the possible strategy which contributes the most towards the minimization of the collection of system objectives. Such a collection of functions is computationally expensive to minimize and also may lead to local minima. In order to avoid this difficulty, the collection of system objectives is deformed into another topological space forming the Homotopy function parameterized by computational temperature T. Due to its analogy to Helmholtz free energy, the approach of Deterministic Annealing (DA) is applied in minimizing the Homotopy function (See Appendix A for details). At every successive temperature drop, the minimization of the Homotopy function is carried out using a second order optimization scheme such as the Nearest Newton Descent Scheme (See Appendix B for details) or Broyden–Fletcher–Goldfarb–Shanno (BFGS) Scheme (See Appendix C for details).

At the end of every iteration, each agent i converges to a probability distribution clearly distinguishing the contribution of its every corresponding strategy value. For every agent, the strategy value with the maximum probability value is referred to as the favorable strategy and is used to compute the system objective. The solution is accepted only if the associated system objective and corresponding strategy values are not worse than the previous iterations solution. In this way, the algorithm continues until convergence by selecting the samples from the neighborhood of the recent favorable strategies.

In some of the applications, the agents are also needed to provide the knowledge of the inter-agent-relationship. It is one of the information/strategy sets which every other entitled agent is supposed to know. There is also global information that every agent is supposed to know. This allows agents to know the right to model other agents' actions or behavior. All of the decisions are taken autonomously by each agent considering the available information in order to optimize the local goals and hence to achieve the optimum global goal or system objective.

The following section describes the detailed formulation of the PC procedure as an unconstrained approach [25–28].

2.2.1 Formulation of Unconstrained PC

Consider a general unconstrained problem (in minimization sense) comprising N variables and objective function G. In the context of PC, the variables of the problem are considered as computational agents/players of a social game being played iteratively.

Each variable, i.e. each agent i is given a predefined sampling interval referred to as $\Psi_i \in \left[\Psi_i^{lower}, \Psi_i^{upper}\right]$. As a general case, the interval can also be referred to as the sampling space. The lower limit Ψ_i^{lower} and upper limit Ψ_i^{upper} of the interval Ψ_i may be updated iteratively as the algorithm progresses.

Each agent i randomly samples $X_i^{[r]}$, $r = 1, 2, \ldots, m_i$ strategies from within the corresponding sampling interval Ψ_i forming a strategy set $\mathbf{X}_i$ represented as

$$\mathrm{X}_i = \{X_i^{[1]}, X_i^{[2]}, X_i^{[3]}, \ldots, X_i^{[m_i]}\}, \quad i = 1, 2, \ldots, N \tag{2.1}$$

Every agent is assumed to have an equal number of strategies, i.e. $m_1 = m_2 = \ldots = m_i = \ldots = m_{N-1} = m_N$. The procedure of modified PC theory is explained below in detail with the algorithm flowchart in Fig. 2.1.

The procedure begins with the initialization of the parameters such as sampling interval Ψ_i for each agent i, temperature $T \gg 0$ or $T = T_{initial}$ or $T \to \infty$ (simply high enough), the temperature step size $\alpha_T (0 < \alpha_T \leq 1)$, convergence parameter $\varepsilon = 0.0001$, interval factor λ_{down} and algorithm iteration counter $n = 1$. The value of α_T, m_i and λ_{down} can be chosen based on preliminary trials of the algorithm.

Step 1 Agent i selects its first strategy $X_i^{[1]}$. Agent i then samples randomly from other agents' strategies as well. These are random guesses by agent i about which strategies have been chosen by the other agents. All these form a 'combined strategy set' $\mathbf{Y}_i^{[1]}$ given by

$$\mathbf{Y}_i^{[1]} = \left\{X_1^{[?]}, X_2^{[?]}, \ldots, X_i^{[1]}, \ldots, X_{N-1}^{[?]}, X_N^{[?]}\right\} \tag{2.2}$$

The superscript [?] indicates that it is a 'random guess' and not known in advance. In addition, agent i forms one combined strategy set for each of the remaining strategies of its strategy set $\mathbf{X}_i$ (i.e. $r = 2, 3, \ldots, m_i$), as shown below.

$$\begin{aligned}
\mathbf{Y}_i^{[2]} &= \left\{X_1^{[?]}, X_2^{[?]}, \ldots, X_i^{[2]}, \ldots, X_{N-1}^{[?]}, X_N^{[?]}\right\} \\
\mathbf{Y}_i^{[3]} &= \left\{X_1^{[?]}, X_2^{[?]}, \ldots, X_i^{[3]}, \ldots, X_{N-1}^{[?]}, X_N^{[?]}\right\} \\
&\vdots \\
\mathbf{Y}_i^{[r]} &= \left\{X_1^{[?]}, X_2^{[?]}, \ldots, X_i^{[r]}, \ldots, X_{N-1}^{[?]}, X_N^{[?]}\right\} \\
&\vdots \\
\mathbf{Y}_i^{[m_i]} &= \left\{X_1^{[?]}, X_2^{[?]}, \ldots, X_i^{[m_i]}, \ldots, X_{N-1}^{[?]}, X_N^{[?]}\right\}
\end{aligned} \tag{2.3}$$

Similarly, all the remaining agents form their combined strategy sets. Furthermore, every agent i computes m_i associated objective function values as follows:

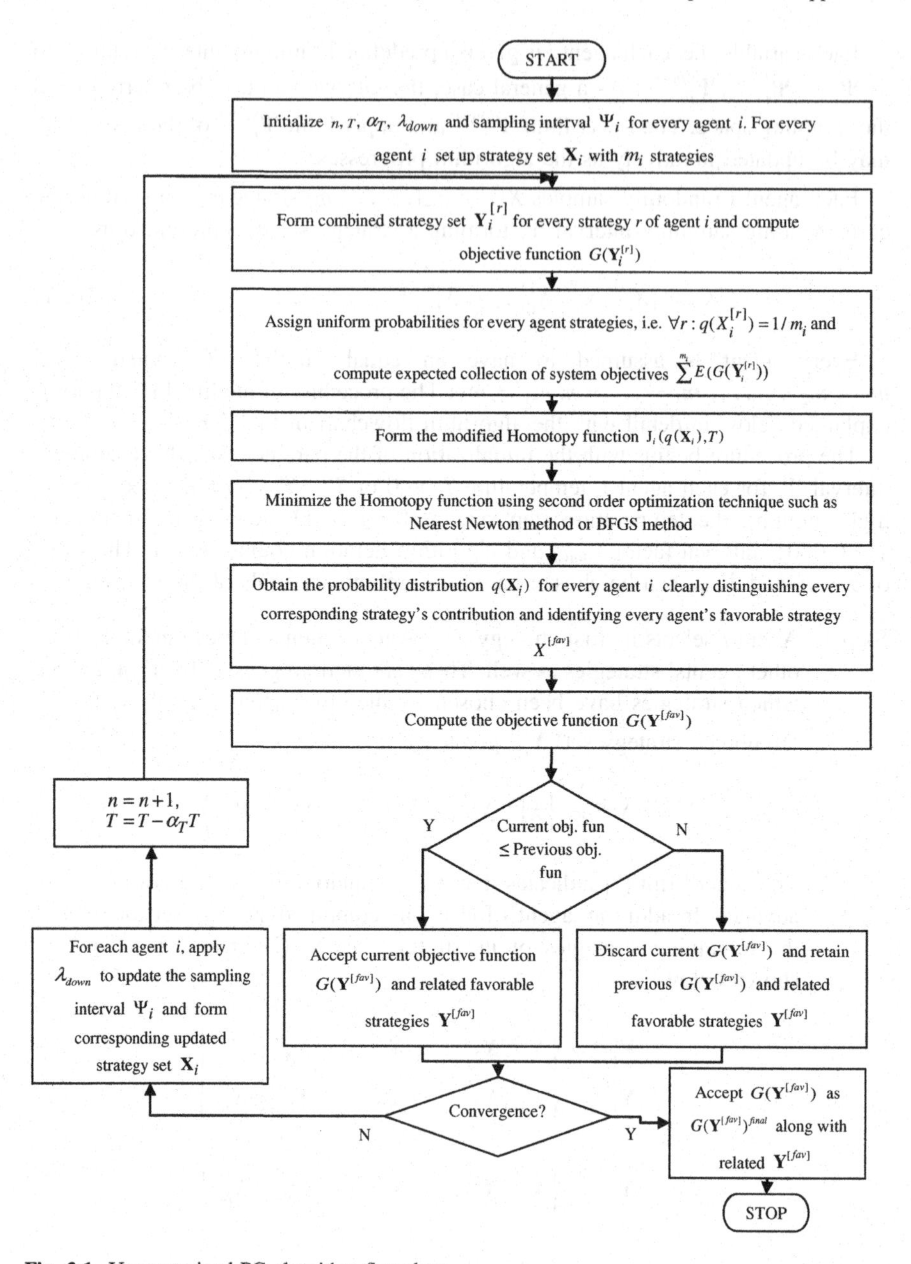

Fig. 2.1 Unconstrained PC algorithm flowchart

$$\left[G\left(\mathbf{Y}_i^{[1]}\right), G\left(\mathbf{Y}_i^{[2]}\right), \ldots, G\left(\mathbf{Y}_i^{[r]}\right), \ldots, G\left(\mathbf{Y}_i^{[m_i]}\right)\right] \tag{2.4}$$

The ultimate goal of every agent i is to identify its strategy value which contributes the most towards the minimization of the sum of these system objective values, i.e. $\sum_{r=1}^{m_i} G(\mathbf{Y}_i^{[r]})$, hereafter referred to as the collection of system objectives. The combined strategy sets, associated objective functions and the collection of system objectives for all the N agents are as follows:

$$\left.\begin{array}{l}
\mathbf{Y}_1^{[1]} = \left\{X_1^{[1]}, X_2^{[?]}, \ldots, X_i^{[?]}, \ldots, X_{N-1}^{[?]}, X_N^{[?]}\right\} \Rightarrow G\left(\mathbf{Y}_1^{[1]}\right) \\
\mathbf{Y}_1^{[2]} = \left\{X_1^{[2]}, X_2^{[?]}, \ldots, X_i^{[?]}, \ldots, X_{N-1}^{[?]}, X_N^{[?]}\right\} \Rightarrow G\left(\mathbf{Y}_1^{[2]}\right) \\
\qquad\vdots \\
\mathbf{Y}_1^{[r]} = \left\{X_1^{[r]}, X_2^{[?]}, \ldots, X_i^{[?]}, \ldots, X_{N-1}^{[?]}, X_N^{[?]}\right\} \Rightarrow G\left(\mathbf{Y}_1^{[r]}\right) \\
\qquad\vdots \\
\mathbf{Y}_1^{[m_i]} = \left\{X_1^{[m_i]}, X_2^{[?]}, \ldots, X_i^{[?]}, \ldots, X_{N-1}^{[?]}, X_N^{[?]}\right\} \Rightarrow G\left(\mathbf{Y}_1^{[m_i]}\right)
\end{array}\right\} \Rightarrow \sum_{r=1}^{m_i} G(\mathbf{Y}_1^{[r]})$$

$$\vdots$$

$$\left.\begin{array}{l}
\mathbf{Y}_i^{[1]} = \left\{X_1^{[?]}, X_2^{[?]}, \ldots, X_i^{[1]}, \ldots, X_{N-1}^{[?]}, X_N^{[?]}\right\} \Rightarrow G\left(\mathbf{Y}_i^{[1]}\right) \\
\mathbf{Y}_i^{[2]} = \left\{X_1^{[?]}, X_2^{[?]}, \ldots, X_i^{[2]}, \ldots, X_{N-1}^{[?]}, X_N^{[?]}\right\} \Rightarrow G\left(\mathbf{Y}_i^{[2]}\right) \\
\qquad\vdots \\
\mathbf{Y}_i^{[r]} = \left\{X_1^{[?]}, X_2^{[?]}, \ldots, X_i^{[r]}, \ldots, X_{N-1}^{[?]}, X_N^{[?]}\right\} \Rightarrow G\left(\mathbf{Y}_i^{[r]}\right) \\
\qquad\vdots \\
\mathbf{Y}_i^{[m_i]} = \left\{X_1^{[?]}, X_2^{[?]}, \ldots, X_i^{[m_i]}, \ldots, X_{N-1}^{[?]}, X_N^{[?]}\right\} \Rightarrow G\left(\mathbf{Y}_i^{[m_i]}\right)
\end{array}\right\} \Rightarrow \sum_{r=1}^{m_i} G(\mathbf{Y}_i^{[r]})$$

$$\vdots$$

$$\left.\begin{array}{l}
\mathbf{Y}_N^{[1]} = \left\{X_1^{[?]}, X_2^{[?]}, \ldots, X_i^{[?]}, \ldots, X_{N-1}^{[?]}, X_N^{[1]}\right\} \Rightarrow G\left(\mathbf{Y}_N^{[1]}\right) \\
\mathbf{Y}_N^{[2]} = \left\{X_1^{[?]}, X_2^{[?]}, \ldots, X_i^{[?]}, \ldots, X_{N-1}^{[?]}, X_N^{[2]}\right\} \Rightarrow G\left(\mathbf{Y}_N^{[2]}\right) \\
\qquad\vdots \\
\mathbf{Y}_N^{[r]} = \left\{X_1^{[?]}, X_2^{[?]}, \ldots, X_i^{[?]}, \ldots, X_{N-1}^{[?]}, X_N^{[r]}\right\} \Rightarrow G\left(\mathbf{Y}_N^{[r]}\right) \\
\qquad\vdots \\
\mathbf{Y}_N^{[m_i]} = \left\{X_1^{[?]}, X_2^{[?]}, \ldots, X_i^{[?]}, \ldots, X_{N-1}^{[?]}, X_N^{[m_i]}\right\} \Rightarrow G\left(\mathbf{Y}_N^{[m_i]}\right)
\end{array}\right\} \Rightarrow \sum_{r=1}^{m_i} G(\mathbf{Y}_N^{[r]}) \tag{2.5}$$

Step 2 For every agent i, the minimum of the function $\sum_{r=1}^{m_i} G(\mathbf{Y}_i^{[r]})$ is very hard to achieve as the function may have many possible local minima. Moreover, directly minimizing this function is quite cumbersome as it may need excessive computational effort. One of the ways to deal with this difficulty is to deform the function into another topological space by constructing a related and 'easier' function E. Such a method is referred to as the Homotopy method. The function E can be referred to as 'easier' because it is easy to compute, the (global) minimum of such a function is known and easy to locate. The deformed function can also be referred to as Homotopy function J parameterized by computational temperature T represented as follows:

$$J_i(q(\mathbf{X}_i), T) = \sum_{r=1}^{m_i} G(\mathbf{Y}_i^{[r]}) - TE \ , \quad T \in [0, \infty) \tag{2.6}$$

(a) Agent i assigns uniform probabilities to its strategies. As an illustration, the uniform probability distribution for agent i may look like that shown in Fig. 2.2 for a case where there are 10 strategies, i.e. $m_i = 10$. This is because, at the beginning, the least information is available (the largest uncertainty and highest entropy) about which strategy is favorable for the minimization of the collection of system objectives $\sum_{r=1}^{m_i} G(\mathbf{Y}_i^{[r]})$. Therefore, at the beginning of the 'game', each agent's every strategy has probability $1/m_i$ of being most favorable. Therefore, probability of strategy r of agent i is

$$q(X_i^{[r]}) = 1/m_i \ , \quad r = 1, 2, \ldots, m_i \tag{2.7}$$

Each agent i, from its every combined strategy set $\mathrm{Y}_i^{[r]}$ and corresponding system objective $G(\mathbf{Y}_i^{[r]})$ computed previously, further computes m_i corresponding expected system objective values $E\left(G(\mathrm{Y}_i^{[r]})\right)$ as follows:

$$E\left(G(\mathrm{Y}_i^{[r]})\right) = G(\mathrm{Y}_i^{[r]})\, q(X_i^{[r]}) \prod_{(i)} q(X_{(i)}^{[?]}) \tag{2.8}$$

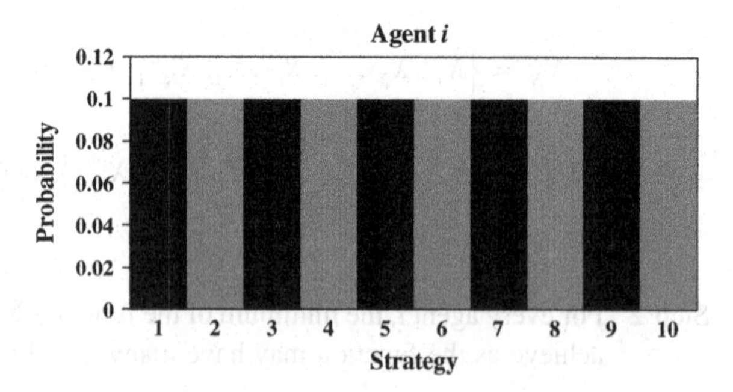

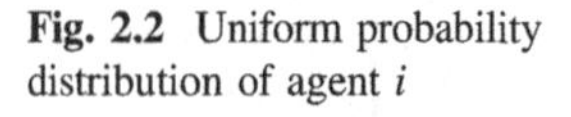
Fig. 2.2 Uniform probability distribution of agent i

where (i) represents every agent other than i. Every agent i then computes the expected collection of system objectives denoted by $\sum_{r=1}^{m_i} E\left(G(\mathbf{Y}_i^{[r]})\right)$. The expected system objectives and the associated expected collection of system objective for all the N agents are as follows:

$$
\left.\begin{array}{l}
G\left(\mathbf{Y}_1^{[1]}\right) q\left(X_1^{[1]}\right) \prod_{(1)} q\left(X_{(1)}^{[?]}\right) = E\left(G\left(\mathbf{Y}_1^{[1]}\right)\right) \\
\vdots \\
G\left(\mathbf{Y}_1^{[r]}\right) q\left(X_1^{[r]}\right) \prod_{(1)} q\left(X_{(1)}^{[?]}\right) = E\left(G\left(\mathbf{Y}_1^{[r]}\right)\right) \\
\vdots \\
G\left(\mathbf{Y}_1^{[m_1]}\right) q\left(X_1^{[m_1]}\right) \prod_{(1)} q\left(X_{(1)}^{[?]}\right) = E\left(G\left(\mathbf{Y}_1^{[m_1]}\right)\right)
\end{array}\right\} \Rightarrow \sum_{r=1}^{m_1} E\left(G(\mathbf{Y}_1^{[r]})\right)
$$

$$\vdots$$

$$
\left.\begin{array}{l}
G\left(\mathbf{Y}_i^{[1]}\right) q\left(X_i^{[1]}\right) \prod_{(i)} q\left(X_{(i)}^{[?]}\right) = E\left(G\left(\mathbf{Y}_i^{[1]}\right)\right) \\
\vdots \\
G\left(\mathbf{Y}_i^{[r]}\right) q\left(X_i^{[r]}\right) \prod_{(i)} q\left(X_{(i)}^{[?]}\right) = E\left(G\left(\mathbf{Y}_i^{[r]}\right)\right) \\
\vdots \\
G\left(\mathbf{Y}_i^{[m_i]}\right) q\left(X_i^{[m_i]}\right) \prod_{(i)} q\left(X_{(i)}^{[?]}\right) = E\left(G\left(\mathbf{Y}_i^{[m_i]}\right)\right)
\end{array}\right\} \Rightarrow \sum_{r=i}^{m_i} E\left(G(\mathbf{Y}_i^{[r]})\right)
$$

$$\vdots$$

$$
\left.\begin{array}{l}
G\left(\mathbf{Y}_N^{[1]}\right) q\left(X_N^{[1]}\right) \prod_{(N)} q\left(X_{(N)}^{[?]}\right) = E\left(G\left(\mathbf{Y}_N^{[1]}\right)\right) \\
\vdots \\
G\left(\mathbf{Y}_N^{[r]}\right) q\left(X_N^{[r]}\right) \prod_{(N)} q\left(X_{(N)}^{[?]}\right) = E\left(G\left(\mathbf{Y}_N^{[r]}\right)\right) \\
\vdots \\
G\left(\mathbf{Y}_N^{[m_N]}\right) q\left(X_N^{[m_N]}\right) \prod_{(N)} q\left(X_{(N)}^{[?]}\right) = E\left(G\left(\mathbf{Y}_N^{[m_N]}\right)\right)
\end{array}\right\} \Rightarrow \sum_{r=N}^{m_N} E\left(G(\mathbf{Y}_N^{[r]})\right)
\tag{2.9}
$$

It also means that the PC approach can convert any discrete variables into continuous variable values in the form of probabilities corresponding to these discrete variables. As mentioned earlier, the problem now becomes continuous but still not easier to solve.

(b) Furthermore, a suitable function for E is chosen. The general choice is to use the entropy function $S_i = -\sum_{r=1}^{m_i} q\left(X_i^{[r]}\right)\log_2 q\left(X_i^{[r]}\right)$. Thus the Homotopy function to be minimized by each agent i in Eq. (2.6) is developed as follows:

$$
\begin{aligned}
J_i(q(\mathbf{X}_i), T) &= \sum_{r=1}^{m_i} E\left(G(\mathbf{Y}_i^{[r]})\right) - T\,S_i \\
&= \sum_{r=1}^{m_i}\left(G(\mathbf{Y}_i^{[r]})\,q(X_i^{[r]})\prod_{(i)} q(X_{(i)}^{[?]})\right) - T\left(-\sum_{r=1}^{m_i} q(X_i^{[r]})\log_2 q(X_i^{[r]})\right) \\
&= G(\mathbf{Y}_i^{[1]})\,q(X_i^{[1]})\prod_{(i)} q(X_{(i)}^{[?]}) + G(\mathbf{Y}_i^{[2]})\,q(X_i^{[2]})\prod_{(i)} q(X_{(i)}^{[?]}) + \cdots \\
&\quad \cdots + G(\mathbf{Y}_i^{[m_i-1]})\,q(X_i^{[m_i-1]})\prod_{(i)} q(X_{(i)}^{[?]}) + G(\mathbf{Y}_i^{[m_i]})\,q(X_i^{[m_i]})\prod_{(i)} q(X_{(i)}^{[?]}) \\
&\quad - T\left(-\sum_{r=1}^{m_i} q(X_i^{[r]})\log_2 q(X_i^{[r]})\right)
\end{aligned}
\tag{2.10}
$$

where $T \in [0, \infty)$.

Step 3 The approach of Deterministic Annealing (DA) discussed in Appendix A is applied to minimize the Homotopy function in Eq. (2.10). The motivation behind this is its analogy to the Helmholtz free energy. Starting from $T \gg 0$ or $T = T_{initial}$ or $T \to \infty$ (simply high enough), with every temperature drop, the minimization of the Homotopy function in Eq. (2.10) can be carried out using suitable optimization technique such as Nearest Newton Descent Scheme as well as a suitable second order optimization approach such as BFGS scheme. The Nearest Newton Descent Scheme and the BFGS scheme minimizing Eq. (2.10) are discussed in Appendix B and Appendix C, respectively

Step 4 For each agent i, the optimization process converges to a probability variable vector $q(\mathbf{X}_i)$ which can be seen as the individual agent's probability distribution distinguishing every strategy's contribution towards the minimization of the expected collection of system objectives $\sum_{r=1}^{m_i} E\left(G(\mathbf{Y}_i^{[r]})\right)$. In other words, for each agent i, if a particular strategy r contributes the most towards the minimization of the objective compared to other strategies, its corresponding probability certainly increases by some amount more than those for the other strategies' probability values,

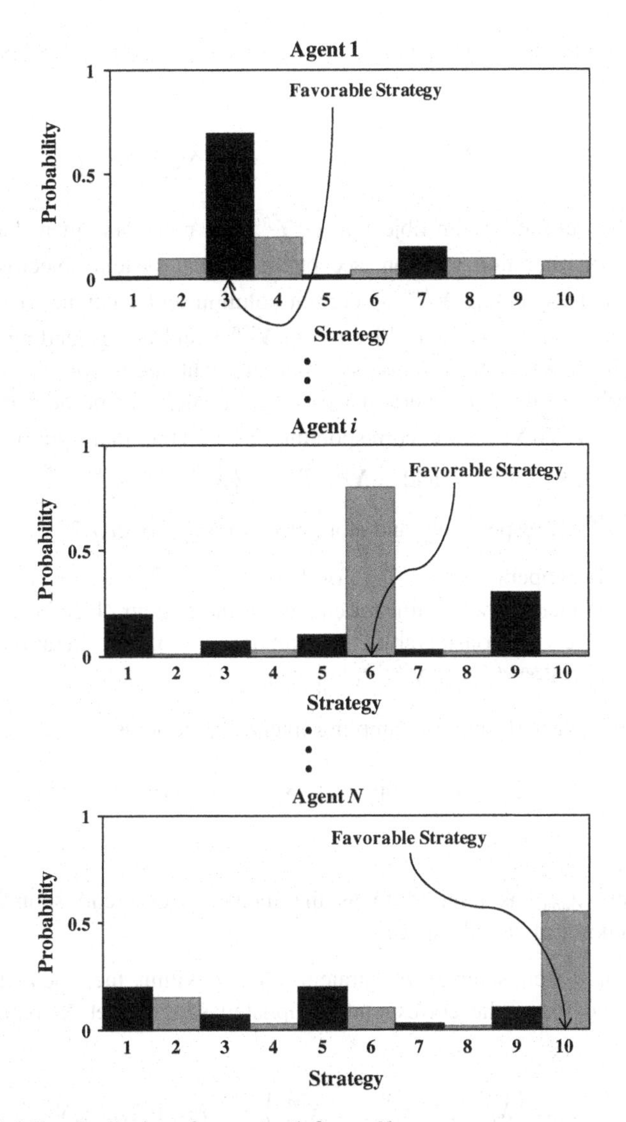

Fig. 2.3 Probability distribution of agent $i = 1, 2, \ldots, N$

and so strategy r is distinguished from the other strategies. Such a strategy is referred to as a favorable strategy $X_i^{[fav]}$. As an illustration, the converged probability distribution for every agent i may look like that shown in Fig. 2.3 for a case where there are 10 strategies, i.e. $m_i = 10$.

The above optimization process when converges, the Nash equilibrium is achieved at the particular temperature T. The concept of Nash equilibrium is discussed in Sect. 2.2.2.

Compute the corresponding system objective $G(\mathbf{Y}^{[fav]})$ where $\mathbf{Y}^{[fav]}$ is given by

$$\mathbf{Y}^{[fav]} = \left\{X_1^{[fav]}, X_2^{[fav]}, \ldots, X_{N-1}^{[fav]}, X_N^{[fav]}\right\} \tag{2.11}$$

Step 5 If the current system objective $G(\mathbf{Y}^{[fav]})$ is not worse than that from the previous iteration solution, accept the current system objective $G(\mathbf{Y}^{[fav]})$ and corresponding $\mathbf{Y}^{[fav]}$ as current solution and continue to *step 6*, else discard current system objective $G(\mathbf{Y}^{[fav]})$ and corresponding $\mathbf{Y}^{[fav]}$, and retain the previous iteration solution and continue to *step 6*.

Step 6 If either of the two criteria listed below is valid, accept the current system objective $G(\mathbf{Y}^{[fav]})$ and corresponding $\mathbf{Y}^{[fav]}$ as the final solution referred to as $G(\mathbf{Y}^{[fav],final})$ and $\mathbf{Y}^{[fav],final} = \left\{X_1^{[fav],final}, X_2^{[fav],final}, \ldots, X_{N-1}^{[fav],final}, X_N^{[fav],final}\right\}$, respectively and stop, else continue to *step 7*.

(a) If temperature $T = T_{final}$ or $T \to 0$.
(b) If there is no significant change in the system objectives for successive considerable number of iterations (i.e. $\left\|G(\mathbf{Y}^{[fav],n}) - G(\mathbf{Y}^{[fav],n-1})\right\| \leq \varepsilon$).

Step 7 Every agent shrinks its sampling interval as follows:

$$\Psi_i \in \left[\left(X_i^{[fav]} - \lambda_{down}\left\|\Psi_i^{upper} - \Psi_i^{lower}\right\|\right), \left(X_i^{[fav]} + \lambda_{down}\left\|\Psi_i^{upper} - \Psi_i^{lower}\right\|\right)\right], 0 < \lambda_{down} \leq 1 \tag{2.12}$$

where λ_{down} is referred to as the interval factor corresponding to the shrinking of sample space.

Each agent i then samples m_i strategies from within the updated sampling interval Ψ_i and forms the corresponding updated strategy set X_i represented as follows.

$$X_i = \left\{X_i^{[1]}, X_i^{[2]}, X_i^{[3]}, \ldots, X_i^{[m_i]}\right\}, \quad i = 1, 2, \ldots, N \tag{2.13}$$

Reduce the temperature $T = T - \alpha_T T$, update the iteration counter $n = n + 1$ and return to *step 1*.

On convergence, the above process reaches the global Nash equilibrium. The concept and detailed formulation of the Nash equilibrium achieved is described in the following section.

2.2.2 Nash Equilibrium in PC

To achieve, a Nash equilibrium, every agent in a MAS should have the properties of Rationality and Convergence [29–32]. Rationality refers to the property by which every agent selects (or converges to) the best possible strategy given the strategies of the other agents. The convergence property refers to the stability condition i.e. a policy using which every agent selects (or converges to) the best possible strategy when all the other agents use their policies from a predefined class (preferably same class). The Nash equilibrium is naturally achieved when all the agents in a MAS are convergent and rational. Moreover, a Nash equilibrium is guaranteed when all the agents use stationary policies, i.e. those policies that do not change over time. It is worth to mention here that all the agents in the MAS proposed using PC algorithm exhibit the above mentioned properties. For example, as detailed in the PC algorithm discussed in Sect. 2.2.1, there is a fixed framework by which the agents select their own strategies and guess the other agents' strategies.

In any game, there may be a large but finite number of Nash equilibria present, depending on the number of strategies per agent as well as the number of agents. It is essential to choose the best possible combination of the individual strategies selected by each agent. It is computationally prohibitive to go through every possible combination of the individual agent strategies and choose the best out of it that can produce a best possible Nash equilibrium and hence the system objective.

As discussed in the detailed PC algorithm, in each iteration n, every agent i selects the best possible strategy referred to as the favorable strategy $X_i^{[fav],n}$ by guessing the possible strategies of the other agents. This information about its favorable strategy $X_i^{[fav],n}$ is made known to all the other agents as well. In addition, the corresponding global knowledge such as system objective value $G(\mathbf{Y}^{[fav],n}) = G\left(X_1^{[fav],n}, X_2^{[fav],n}, X_3^{[fav],n}, \ldots, X_{N-1}^{[fav],n}, X_N^{[fav],n}\right)$ is also available to each agent which clearly helps all the agents take the best possible informed decision in every further iteration. This makes the entire system ignore a considerably large number of Nash equilibria but select the best possible one in each iteration and accept the corresponding system objective $G(\mathbf{Y}^{[fav],n})$. Mathematically the Nash equilibrium solution at any iteration of the PC algorithm can be represented as follows:

$$\begin{gathered}
G\left(X_1^{[fav],n}, X_2^{[fav],n}, \ldots, X_{N-1}^{[fav],n}, X_N^{[fav],n}\right) \le G\left(X_1^{(fav),n}, X_2^{[fav],n}, \ldots, X_{N-1}^{[fav],n}, X_N^{[fav],n}\right) \\
G\left(X_1^{[fav],n}, X_2^{[fav],n}, \ldots, X_{N-1}^{[fav],n}, X_N^{[fav],n}\right) \le G\left(X_1^{[fav],n}, X_2^{(fav),n}, \ldots, X_{N-1}^{[fav],n}, X_N^{[fav],n}\right) \\
\vdots \\
G\left(X_1^{[fav],n}, X_2^{[fav],n}, \ldots, X_{N-1}^{[fav],n}, X_N^{[fav],n}\right) \le G\left(X_1^{[fav],n}, X_2^{[fav],n}, \ldots, X_{N-1}^{[fav],n}, X_N^{(fav),n}\right)
\end{gathered} \tag{2.14}$$

where $X_i^{(fav),n}$ represents any strategy other than the favorable strategy $X_i^{[fav],n}$ from the same sample space Ψ_i^n.

Furthermore, from this current Nash equilibrium point with system objective $G(\mathbf{Y}^{[fav],n})$, the algorithm progresses to the next Nash equilibrium point with better system objective $G(\mathbf{Y}^{[fav],n+1})$, i.e. $G(\mathbf{Y}^{[fav],n}) \geq G(\mathbf{Y}^{[fav],n+1})$. As the algorithm progresses, those ignored Nash equilibria as well as the best Nash equilibria selected at previous iterations would be noticed as inferior solutions.

This process continues until there is no change in the current solution $G(\mathbf{Y}^{[fav],n})$, i.e. no new Nash equilibrium has been identified that proves the current Nash equilibrium to be inferior. Hence the system exhibits stage-wise convergence to a unique Nash equilibrium and the corresponding system objective is accepted as the final solution $G(\mathbf{Y}^{[fav],final})$. As a general case, this progress can be represented as follows:

$$G(\mathbf{Y}^{[fav],1}) \geq G(\mathbf{Y}^{[fav],2}) \geq \cdots \geq G(\mathbf{Y}^{[fav],n}) \geq G(\mathbf{Y}^{[fav],n+1}) \geq \cdots \geq G(\mathbf{Y}^{[fav],final}).$$

2.3 Characteristics of PC

The PC approach has the following key characteristics that make it a competitive choice over other algorithms for optimizing collectives.

1. PC is a distributed solution approach in which each agent independently updates its probability distribution at any time instance and can be applied to continuous, discrete or mixed variables, etc. Since the probability distribution of the strategy set is always a vector of real numbers regardless of the type of variable under consideration, conventional techniques of optimization for Euclidean vectors, such as gradient descent, can be exploited. This is explicitly demonstrated in Chap. 6.
2. It is robust in the sense that the cost function (global/system objective) can be irregular or noisy, i.e. it can accommodate noisy and poorly modeled problems.
3. The failed agent can just be considered as one that does not update its probability distribution, without affecting the other agents. On the other hand, it may severely hamper the performance of other techniques. This is explicitly demonstrated in Chap. 5.
4. It provides the sensitivity information about the problem in the sense that a variable with a peaky distribution (having highest probability value) is more important in the solution than a variable with a broad distribution, i.e. peaky distribution provides the best choice of action that can optimize the global goal. This is explicitly demonstrated in Sect. 2.2.1.
5. The formation of the Homotopy function for each agent (variable) helps the algorithm to jump out of the possible local minima and further reach the global

minima. This is explicitly discussed in Appendix A and also evident when solved a variety of functions using Penalty Function approach in Chap. 4.

6. It can successfully avoid the tragedy of commons, skipping the local minima and further reach the true global minima. This is explicitly demonstrated in Chap. 5.
7. The computational and communication load is significantly less and equally distributed among all the agents.
8. It can efficiently handle problems with moderately large number of variables. This is demonstrated in Chaps. 6 and 7.

With PC solving optimization problems as a MAS, it is worth discussing some of its characteristics to compare the similarities and differences with Multi-Agent Reinforcement Learning (MARL) methods.

9. Most MARL methods such as fully cooperative, fully competitive and mixed (neither cooperative nor competitive) are based on Game Theory, Optimization and Evolutionary Computation. Most of these types of methods possess less scalability and are sensitive to imperfect observations. Any uncertainty or incomplete information may lead to unexpected behavior of the agents. However, the scalability of the fully cooperative methods such as coordination free methods can be enhanced by explicitly using the communication and/or uncertainty techniques. On the other hand, being a distributed optimization technique PC is scalable, and can handle uncertainty in terms of probability. Moreover, the random strategies selected by any agent can be coordinated or negotiated with the other agents based on the social conventions, right to communication, etc. This social aspect makes PC a cooperative approach.
10. Furthermore, indirect coordination based methods work on the concept of biasing the selection towards the likelihood of the good strategies. This concept is similar to the one used in the PC algorithm presented here, in which agents choose the strategy sets only in the neighborhood of the best strategy identified in the previous iteration.
11. In the case of mixed MARL algorithms, the agents have no constraints imposed on their rewards. It is similar to the PC algorithm in which the agents respond or select the strategies and exhibit self-interested behavior. However, the mixed MARL algorithms may encounter multiple Nash Equilibria while in PC a unique Nash equilibrium can be achieved.

2.4 Modified PC Approach Versus Original PC Approach

It is worth to mention some of the key differences of the PC methodology presented here and the original PC approach proposed by Dr. David Wolpert [1–4]

1. In the PC approach presented here, fewer numbers of samples were drawn from the uniform distribution of the individual agent's sampling interval. On the

contrary, the original PC approach used a Monte Carlo sampling method which was computationally expensive and slow as the number of samples needed was in the thousands or even millions.

2. Most significantly, the sampling in further stages of the PC algorithm presented here was narrowed down in every iteration by selecting the sampling interval in the neighborhood of the most favorable value in the particular iteration. This ensures faster convergence and an improvement in efficiency over the original PC approach in which regression was necessary to sample the strategy values in the close neighborhood of the favorable value.
3. Moreover, the coordination among the agents representing the variables in the system is achieved based on the partial small bit of information. In other words, every agent selects its best possible strategy by guessing the model of every other agent based merely on their recent favorable strategies communicated. This gives the advantage to the agents and the entire system to quickly search the better solution and reach the Nash equilibrium.

2.5 Validation of the Unconstrained PC

There are a number of benchmark test functions for contemporary optimization algorithms like GAs and evolutionary computation. The Rosenbrock function (see Fig. 2.4) is an example of a nonlinear function having strongly coupled variables and is a real challenge for any optimization algorithm because of its slow convergence for most optimization methods [33, 34]. The Rosenbrock function with N number of variables is given by

$$G = \sum_{i=1}^{N-1} \left[100\left(x_i^2 - x_{i+1}\right)^2 + (1 - x_i)^2 \right] \tag{2.15}$$

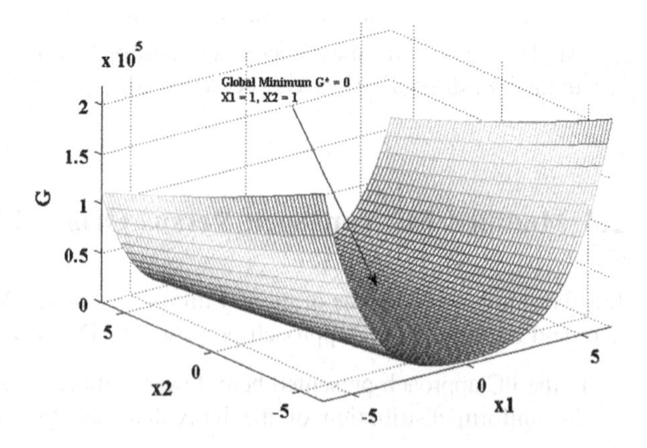

Fig. 2.4 Rosenbrock function

$$\Psi_i \in \left[\Psi_i^{lower}, \Psi_i^{upper}\right], \quad i = 1, 2, \ldots, N$$

and the global minimum is at, $x_i = 1, \forall i \; i = 1, 2, \ldots, N$, where $G^* = 0$. A difficulty arises from the fact that the optimum is located in a deep and narrow parabolic valley with a flat bottom. Also, gradient based methods may have to spend a large number of iterations to reach the global minimum.

In the context of PC, every function variable x_i , $i = 1, 2, \ldots, N$ was represented as an autonomous agent. These agents competed with one another to optimize their individual values and ultimately the entire function value (i.e. global objective G). The Rosenbrock function with 5 variables/agents ($N = 5$) was solved here in which each agent randomly selected the strategies from within a predefined sampling interval Ψ_i. Although the optimal value of each and every agent x_i is 1, the allowable sampling interval Ψ_i for each agent was intentionally assigned to be different (as shown in Table 2.1). The procedure explained in Sect. 2.2.1 was followed to reach convergence. For all the runs, the temperature step size α_T, the interval factor λ_{down} and sample size m_i for each agent chosen were 0.9, 0.1 and 42, respectively. It is worth to mention that the temperature T was reduced on completion of every 10 iterations.

The PC algorithm discussed in Sect. 2.2.1 was coded in MATLAB 7.4.0 (R2007A) on Windows platform using a Pentium 4, 3 GHz processor speed and 512 MB RAM. The results of 5 runs are shown in Table 2.1. The convergence plot for run 5 is shown in Fig. 2.5. For run 5, the value of the function G and associated variable values at iteration 113 was accepted as the final solution as there was no change in the solution for a considerable number of iterations. The variations in the function value and the number of function evaluations among the runs are due to the randomness in selection of the strategies.

A number of researchers have solved the Rosenbrock function using various algorithms. Table 2.2 summarizes the results obtained by these various algorithms:

Table 2.1 Performance using PC approach

Agents or (variables)	Strategy values selected with maximum probability					
	Run 1	Run 2	Run 3	Run 4	Run 5	Range of values (Ψ_i)
Agent 1	1	0.9999	1.0002	1.0001	0.9997	−1.0 to 1.0
Agent 2	1	0.9998	1.0001	1.0001	0.9994	−5.0 to 5.0
Agent 3	1.0001	0.9998	1	0.9999	0.9986	−3.0 to 3.0
Agent 4	0.9998	0.9998	0.9998	0.9995	0.9967	−3.0 to 8.0
Agent 5	0.9998	0.9999	0.9998	0.9992	0.9937	1.0 to 10.0
Fun. value G	2×10^{-5}	1×10^{-5}	2×10^{-5}	2×10^{-5}	5×10^{-5}	
Fun. evaluations	288,100	223,600	359,050	204,750	249,500	

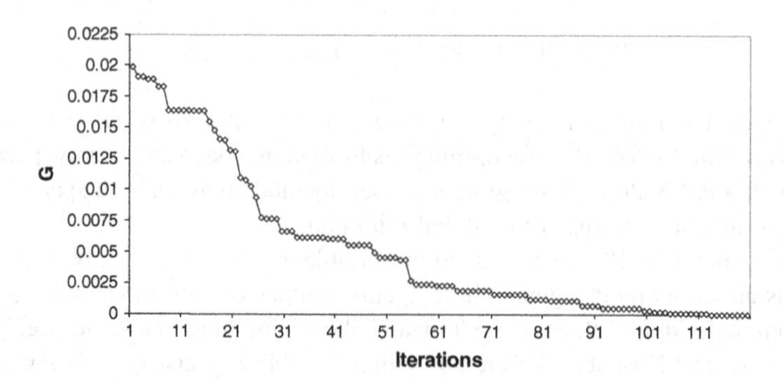

Fig. 2.5 Convergence plot for run 5

Chaos Genetic Algorithm (CGA) [33], Punctuated Anytime Learning (PAL) system [34], Modified Differential Evolution (Modified DE) proposed in [35] and Loosely Coupled GA (LCGA) implemented in [36] are some of the algorithms demonstrated on the Rosenbrock function.

Within each of the results shown in Table 2.2, every variable was assigned an identical interval of allowable values, different from Table 2.1 where each variable was assigned a different allowable interval. Furthermore, even with the larger number of variables, the optimal function values in Table 2.1 are better than those in Table 2.2 except for the modified DE results. This makes it clear that the approach presented here produced fairly good comparable results to those produced by previous researchers. In addition, several unconstrained test problems were solved and the solutions (refer to Table 2.3) achieved were very close to the true optimum solution with acceptable computation cost.

The following few chapters discuss various constraint handling techniques incorporated into the PC which made it a more generic and powerful optimization technique.

Table 2.2 Performance comparison of various algorithms solving Rosenbrock function

Method	Number of agents/ variables (N)	Function value G	Function evaluations	Range of values (Ψ_i)
CGA [33]	2	0.000145	250	−2.048 to 2.048
PAL [34]	2	≈0.01	5,250	−2.048 to 2.048
	5	≈2.5	100,000	−2.048 to 2.048
Modified DE [35]	2	1×10^{-6}	1,089	−5 to 10
	5	1×10^{-6}	11,413	−5 to 10
LCGA [36]	2	≈0.00003	–	−2.12 to 2.12

Table 2.3 Other unconstrained test problems

Function	True optimum solution	PC solution	Average function evaluations	Average CPU time (s)
Ackley	0	0.0000945	23,665	212.40
Beale	0	0	55	0.48
Bohachevsky	0	0	179,747	61.14
Booth	0	0	188,657	100.28
Branin	0.397887	0.397	128,079	109.69
Colville	0	0.00007	612,208	296.65
Rosenbrock	0	0	248,223	179.63
Zakharov	0	1.19E-10	197,550	125.45
Sum squares	0	5.79E-11	176,840	70.75
Shubert	−186.7309	−186.7096	234,734	121.85
Rastrigin		0.00000132	28,709	1,152
Power sum	0	0	221	0.05
Powell	0	0.000000196	774,556	310.65
Perm	0	9.4E-09	162,333	167.65
Michalewics	−1.8013	−1.8013	4,482	2.65
Matyas	0	0	191,898	127.15
Hump	0	0	96,119	51.07
Dixon and price	0	0.00000692	539,267	186.46
Goldstein and price	3	3	169,364	384.82
Griewank	0	1.79E-12	227,517	123.09
Hartmann (3 variables)	−3.86278	−3.86	133,038	70.03
Hartmann (6 variables)	−3.32237	−3.3223	31,535	21.45
Levy	0	0.0000041	161,394	85.78
Shifted sphere	−450	−450	42,653	13.76
Shifted rastrigin	−330	−330	36,032	14.28
Shifted ackley	−140	−140	213,296	276.75
Shifted greiwank	−180	−180	1,041	1.31
Schwefel	0	0	143,732	77.60

References

1. Wolpert, D.H., Tumer, K.: An introduction to collective intelligence. Technical Report, NASA ARC-IC-99–63, NASA Ames Research Center (1999)
2. Bieniawski, S.R.: Distributed optimization and flight control using collectives. Ph.D dissertation, Stanford University, CA, USA, (2005)
3. Wolpert, D.H.: Information theory—the bridge connecting bounded rational game theory and statistical physics. In: Braha, D., Minai, A.A., Bar-Yam, Y. (eds.) Complex Engineered Systems, pp. 262–290. Springer (2006)
4. Wolpert, D.H., Strauss, C.M.E., Rajnarayan, D.: Advances in distributed optimization using probability collectives. Adv. Complex Syst. **9**(4), 383–436 (2006)
5. Wolpert, D.H., Antoine, N.E., Bieniawski, S.R., Kroo, I.R.: Fleet assignment using collective intelligence. In: Proceedings of the 42nd AIAA Aerospace Science Meeting Exhibit (2004)
6. Bieniawski, S.R., Kroo, I.M., Wolpert, D.H.: Discrete, continuous, and constrained optimization using collectives. In: 10th AIAA/ISSMO Multidisciplinary Analysis and Optimization Conference, vol. 5, pp. 3079–3087 (2004)
7. Huang, C.F., Chang, B.R.: Probability collectives multi-agent systems: a study of robustness in search. LNAI 6422, Part II, pp. 334–343 (2010)
8. Huang, C.F., Bieniawski, S., Wolpert, D., Strauss, C.E.M.: A comparative study of probability collectives based multiagent systems and genetic algorithms. In: Proceedings of the Conference on Genetic and Evolutionary Computation, pp. 751–752 (2005)
9. Luo, D.L., Shen, C.L., Wang, B., Wu, W.H.: Air combat decision-making for cooperative multiple target attack using heuristic adaptive genetic algorithm. In: Proceedings of IEEE International Conference on Machine Learning and Cybernetics IEEE Press, pp. 473–478 (2005)
10. Luo, D.L., Duan, H.B., Wu, S.X., Li, M.Q.: Research on air combat decision-making for cooperative multiple target attack using heuristic ant colony algorithm. Acta Aeronautica et Astronautica Sinica **27**(6), 1166–1170 (2006)
11. Luo, D.L., Yang, Z., Duan, H.B., Wu, Z.G., Shen, C.L.: Heuristic particle swarm optimization algorithm for air combat decision-making on CMTA. Trans. Nanjing Univ. Aeronaut. Astronaut. **23**(1), 20–26 (2006)
12. Zhang, X.P, Yu, W.H., Liang, J.J., Liu, B.: Entropy regularization for coordinated target assignment. In: Proceedings of 3rd IEEE Conference on Computer Science and Information Technology, pp. 165–169 (2010)
13. Vasirani, M., Ossowski, S.: Collective-based multiagent coordination: a case study. LNAI **4995**, 240–253 (2008)
14. Modi, P., Shen, W., Tambe, M., Yokoo, M.: Adopt: asynchrous distributed constraint optimization with quality guarantees. Artif. Intell. **161**, 149–180 (2005)
15. Mohammad, H.A., Babak, H.K.: A distributed probability collectives optimization method for multicast in CDMA wireless data networks. In: Proceedings of 4th IEEE International Symposium on Wireless Communication Systems, art. No. 4392414, pp. 617–621 (2007)
16. Ryder, G.S., Ross, K.G.: A probability collectives approach to weighted clustering algorithms for ad hoc networks. In: Proceedings of Third IASTED International Conference on Communications and Computer Networks, pp. 94–99 (2005)
17. Goldberg, D.E., Samtani, M.P.: Engineering optimization via genetic algorithm. In: Proceedings of 9th Conference on Electronic Computation, pp. 471–484 (1986)
18. Ghasemi, M.R., Hinton, E., Wood, R.D.: Optimization of trusses using genetic algorithms for discrete and continuous variables. Eng. Comput. **16**(3), 272–301 (1999)
19. Moh, J., Chiang, D.: Improved simulated annealing search for structural optimization. AIAA J. **38**(10), 1965–1973 (2000)
20. Autry, B.: University course timetabling with probability collectives. Master's thesis, Naval Postgraduate School Montery, CA, USA (2008)

21. Sislak, D., Volf, P., Pechoucek, M., Suri, N.: Automated conflict resolution utilizing probability collectives optimizer. IEEE Trans. Syst. Man Cybern.: Appl. Rev. **41**(3), 365–375 (2011)
22. Arora, J.S.: Introduction to Optimum Design. Elsevier Academic Press (2004)
23. Vanderplaat, G.N.: Numerical Optimization Techniques for Engineering Design. Mcgraw-Hill, New York (1984)
24. Smyrnakis, M., Leslie, D.S.: Sequentially updated probability collectives. In: Proceedings of 48th IEEE Conference on Decision and Control and 28th Chinese Control Conference, pp. 5774–5779 (2009)
25. Kulkarni, A.J., Tai, K.: Probability collectives for decentralized, distributed optimization: a collective intelligence approach. In: Proceedings of the IEEE International Conference on Systems, Man, and Cybernetics, pp. 1271–1275 (2008)
26. Kulkarni, A.J. Tai, K.: Probability collectives: a decentralized, distributed optimization for multi-agent systems. In: Mehnen, J., Koeppen, M., Saad, A., Tiwari, A. (eds.) Applications of Soft Computing, pp. 441–450. Springer (2009)
27. Kulkarni, A.J., Tai, K.: Solving constrained optimization problems using probability collectives and a penalty function approach. Int. J. Comput. Intell. Appl. **10**(4), 445–470 (2011)
28. Kulkarni A.J., Tai, K.: A probability collectives approach with a feasibility-based rule for constrained optimization. Appl. Comput. Intell. Soft Comput. 2011, Article ID 980216
29. Shoham, Y., Powers, R., Grenager, T.: Multi-agent reinforcement learning: a critical survey. www.cc.gatech.edu/~isbell/reading/papers/MALearning.pdf Accessed 23 July 2011
30. Busoniu, L., Babuska, L., Schutter, B.: A comprehensive survey of multiagent reinforcement learning. IEEE Trans. Syst Man Cybern.—Part C: Appl. Rev. **38**(2), 156–172 (2008)
31. Bowling, M., Veloso, M.: Multiagent learning using a variable learning rate. Artif. Intell. **136** (2), 215–250 (2002)
32. Bowling, M., Veloso, M.: Rational and convergent learning in stochastic games. In: Proceedings of 17th International Conference on Artificial Intelligence, pp. 1021–1026 (2001)
33. Cheng, C.T., Wang, W.C., Xu, D.M., Chau, K.W.: Optimizing hydropower reservoir operation using hybrid genetic algorithm and chaos. Water Resour. Manag. **22**, 895–909 (2008)
34. Blumenthal, H.J., Parker, G.B.: Benchmarking punctuated anytime learning for evolving a multi-agent team's binary controllers. In: Proceedings of World Automation Congress, pp. 1–8 (2006)
35. Roger, L.S., Tan, M.S., Rangaiah, G.P.: Global optimization of benchmark and phase equilibrium problems using differential evolution. http://www.ies.org.sg/journal/current/v46/v462_3.pdf
36. Bouvry, P., Arbab, F., Seredynski, F.: Distributed evolutionary optimization, in manifold: rosenbrock's function case study. Inf. Sci. **122**, 141–159 (2000)

Chapter 3
Constrained Probability Collectives: A Heuristic Approach

This chapter discusses an approach of handling the constraints in which the problem specific information is explicitly used. Basically, this approach involves development of problem specific heuristic techniques and further combines them with the unconstrained optimization technique to push the objective function into the feasible region.

The approach of PC was successfully combined with several problem specific heuristic techniques to handle the constraints. The approach was validated by solving two test cases of the Multiple Depot Multiple Traveling Salesmen Problem (MDMTSP) [1] as well as several cases of the Single Depot MTSP (SDMTSP) [2]. In these cases the vehicles are considered as autonomous agents collectively searching the minimum cost path.

The following few sections describe the MTSP and Vehicle Routing Problem (VRP) in detail including the possible extension to Multiple Unmanned Aerial Vehicles (MUAVs). It is followed by the detailed explanation of the proposed MTSP specific heuristic techniques solving various cases of the MDMTSP and SDMTSP. The results are discussed at the end of the chapter.

3.1 The Traveling Salesman Problem (TSP)

The Traveling Salesman Problem (TSP) represents an important class of optimization problems also referred to as NP-hard. Although very easy to understand it is very hard to solve. This is because as the number of cities (n) increases, the number of possible routes increases exponentially. According to Durbin and Willshaw [3] and Papadimitriou [4], although these problems are inter-convertible, the computational complexity increases faster than any power of n. The inter-convertibility refers to an important characteristic that if a resource efficient solution to any one of them is discovered, it can be adapted to solve other NP-complete problems.

© Springer International Publishing Switzerland 2015

A.J. Kulkarni et al., *Probability Collectives*, Intelligent Systems Reference Library 86, DOI 10.1007/978-3-319-16000-9_3

3.1.1 The Multiple Traveling Salesmen Problem (MTSP)

A generalization of the well known Traveling Salesman Problem (TSP) is the Multiple Traveling Salesman Problem (MTSP). As the name indicates, there is more than one traveling salesman in the MTSP. The different variants of the MTSP seem more appropriate for real-life applications [5]. The literature on TSP and MTSP suggests that as compared to the former, the latter received little attention from researchers. The generalized representation of MTSP is as follows.

Consider a complete directed graph $G = (\mathbf{N}, \mathbf{E})$, where $\mathbf{N} = \{1, 2, \ldots, n\}$ is the set of customer nodes or vertices or cities to be visited and $\mathbf{E}$ is the set of arcs connecting the customer nodes. The number of salesmen located at the depot is m. Generally, node 1 is considered as the depot which is the starting and end point of the salesmen's journey. All the salesmen start from the depot, visit the remaining nodes exactly once and return to the same depot. There is a non-negative cost c_{ij} associated with each arc connecting the nodes i and j which may represent the distance to travel or time to travel. Generally, the problems are symmetrical, hence $c_{ij} = c_{ji}$ for every set of arc (i, j). While following these routes, the total cost of visiting all the nodes is to be minimized. Generally, the cost metric is defined in terms of the total distance or the total time to travel the distance between the nodes, etc. [6–9].

The applications of MTSP include the areas of logistics and manufacturing scheduling [10, 11], delivery and pick-up scheduling [12, 13], school bus delivery routing [14–16], satellite navigation systems [17], problems like overnight security problem [18], etc.

3.1.1.1 Variations of the MTSP

Depending on the applications and requirements, different constraints are added to the general MTSP resulting in different variations of the problem. These are discussed below.

The first variation to that is a single/multiple depot case in which all the salesmen return to their original depot/s at the end of their tours. This case is referred to as a fixed destination case [5].

The second variation is also a multiple depot case but it is not necessary that all the salesmen return to their original depots at the end of their tours but there is a constraint that all the depots should have the same number of salesmen at the end of all travels as at the beginning. This case is referred to as non-fixed destination case [5].

The third variation is that "the number of salesmen in the problem may be a bounded variable or fixed a priori" [5]. It is associated with assigning a fleet of vehicles to the predefined routes. Based on the demand, the fleet size may vary within the predefined limits. This has application to the transportation planning such as school bus routing, local truckload pick-up and delivery, as well as the satellite surveying system.

According to Bektas [5], the fourth variation is when the number of salesmen in the problem is not fixed. So whenever such salesmen are to be used, there is an associated cost with each salesman which is to be minimized by minimizing the number of salesmen in the solution.

The next variation is referred to as MTSP with Time Window (MTSPTW). This refers to the problem having the constraint that certain nodes need to be visited within specific period of time. The MTSP application to school-bus and airplane/ship fleet scheduling are characterized with such cases. MTSPTW has received attention recently for applications such as security crew scheduling. In MTSPTW, there is a Time Window (earliest start time, latest end time) applied to visit any customer node. The traveling salesman is allowed to visit the node within this Time Window only. If he arrives earlier than the 'earliest start time' he has to wait. If he arrives after 'latest end time' he will be penalized. The Time Window (earliest start time, latest end time) is also referred to as (lower bound, upper bound) [19]. The difference between the 'earliest start time' and the 'latest end time' is generally referred to as the service time. Sometimes there is only the upper bound associated with each customer node, then such MTSP can be referred to as MTSP with Time Deadlines (MTSPTD).

The Sixth variation is the restriction on the number of nodes each salesman visits. In addition to that, there can be an upper and lower limit on the distance traveled by a salesman.

3.1.2 The Vehicle Routing Problem (VRP)

The VRP is the extension to MTSP in which the cost of travel is minimized by considering traveling vehicles' capacity and the demand associated with each customer node.

Due to its economic importance in various distribution applications, VRP has received growing attention. For the general VRP, the routes are designed so that

1. Each demand location is served exactly once
2. The total demand on each route is less than or equal to the capacity of the vehicle assigned to that route
3. It has single depot (multiple depot in a few cases), i.e. all the vehicles start and complete the route at a single central depot.

In this sense, MTSP and VRP are closely related. In VRP there is capacity constraint which limits the amount of service demanded on a route. On the other hand, relaxing this constraint by giving unlimited capacity to the vehicles will turn the VRP into a MTSP. The generalized representation of VRP is as follows:

Let $G = (\mathbf{N}, \mathbf{E})$ be a graph, where $\mathbf{N} = \{1, 2, \ldots, n\}$ is the customer nodes set and $\mathbf{E}$ is the connecting arc set. Node 1 is the depot which is the starting and end point of the fleet of vehicles. Q is the capacity of an individual vehicle and in general, it is the same for all the vehicles. q_i is the demand at each customer node $i \in \mathbf{N} - \{1\}$. Similar to the MTSP, there is a non-negative cost c_{ij} associated with each arc connecting the nodes i and j which may represent the distance to travel or time to travel. Generally the problems are symmetrical, hence $c_{ij} = c_{ji}$ for every set of arc (i, j).

VRP has real-world applications in the manufacturing sector, including supply chains involving the delivery of goods from suppliers to factories and from factories to customers [19]. Some of the application areas according to [20] are the delivery of meals [21], packaged foods [22], chemical products [23], delivery of gas [24], etc. In such applications, the minimization of the number of vehicles is also important. The garbage collection, mail delivery and public transportation are also some of the application areas in the service sector [19].

3.1.2.1 Variations of the VRP

Similar to the MTSP, there are some variations to the general VRP definition. As mentioned previously, these variants depend on the applications but more on the types of constraints to be satisfied. Potvin [19] provided a detailed discussion on VRP variations.

Similar to the MTSP, there are variations to the VRP known as VRP with Time Window (VRPTW) and VRP with Time Deadlines (VRPTD).

There is a real world variation to VRP known as the Time-Dependent Vehicle Routing Problem (TDVRP) [19]. In this case, the traveling distance is fixed but the time is not fixed. This is because of peak (may take longer to complete a route) and non-peak (may be quicker as compared to peak hours) traffic hours.

Pick-up and Delivery Problem (PDP) is also a variation to the general VRP such that the corresponding pair of nodes should be on the same route. In this pair of nodes, one is a pick-up node ($i-$) and another is a delivery node ($i+$). So now this VRP is associated with two constraints. The first one is that the pick-up and delivery node should be on the same route. The second one is that delivery node should follow pick-up node. The former constraint is referred to as a 'pairing constraint' while the latter is referred to as a 'precedence constraint'. When Time Window is considered, the PDP is referred to as PDPTW [19].

Another real world variation is Vehicle Routing Problem with Backhauls (VRPB). In this variation, the customer nodes are partitioned into linehaul and backhaul nodes. The linehaul nodes are the ones which are to be visited during the forward journey i.e. demand is to be delivered to these nodes, while the backhaul nodes are the ones from which the demand is brought back during the return journey to the depot. The precedence constraint has to be considered in this case. When Time Window is considered, the VRPB can be referred to as VRPBTW. According to the analysis by Ghaziri and Osman [25], the VRPB can be solved exactly for smaller instances and approximately for the larger sized problems.

In some of the cases where the fleet of vehicles is very limited or the length of the day is long as compared to the average duration of the routes, the VRP with Multiple Depots (MDVRP) is used. It is also referred to as Vehicle Routing Problem with Multiple Use of Vehicles (VRPM) [26].

A variation to the MDVRP is where the depots are considered as intermediate replenishment stations along the routes of the vehicles. Such MDVRP is referred to as Multi-Depot Vehicle Routing Problem with Inter-Depot Routes (MDVRPI) [27]. Applications such as the truck and trailer (leaving a trailer at a depot and carrying one for the next journey) can be solved using MDVRPI.

Potvin [19] discussed some short details about other VRP variations such as multi-depot, mixed fleet and size, periodic vehicle routing, location-routing, inventory routing, stochastic vehicle routing, and dynamic vehicle routing problems.

3.1.3 Algorithms and Local Improvement Techniques for Solving the MTSP

There are various bio-inspired optimization algorithms such as GA, Ant Colony Optimization (ACO), Artificial Neural Networks (ANN), Particle Swarm Optimization (PSO) used to solve the variants of the TSP/MTSP/VRP. These algorithms are used in conjunction with local improvement techniques to jump out of local minima and also to reduce the computational load and time. These algorithms are discussed, highlighting the various weaknesses when solving the MTSP/VRP. The respective importance of the local improvement techniques is also discussed. In the PC work presented here, efforts were taken to best exploit some of these local improvement techniques and also to overcome some of the weaknesses of the other algorithms.

3.1.3.1 Local Improvement Techniques

The commonly used local improvement techniques are insertion, savings principle, swapping of nodes, Simulated Annealing (SA), r-opt technique, etc. The insertion and the savings principle are generally applied in forming the routes while the swapping and r-opt techniques are applied for further improvement on forming of the feasible routes. In the insertion technique, the nodes are iteratively assigned to the vehicles to form the feasible routes. Once the feasible routes are formed, the above mentioned optimization techniques are applied to further optimize the routes. The savings principle is based on the traveling cost between the depot and two nodes d_{0i} and d_{0j} (where 0 is the depot, i and j are the nodes) and the traveling cost between the two nodes d_{ij}. The saving is computed as $S_{ij} = d_{0i} + d_{0j} - d_{ij}$. The savings principle ranks these savings in descending order, helping to cluster the closer nodes. The insertion and savings principle help to reduce the computational overload when forming the routes. These replace the possible exhaustive approach of using permutations to form the routes and selecting the best from them.

The nodes belonging to two routes are exchanged in the swapping technique which is a strong local improvement technique. Chen et al. [28] demonstrated that the SA does the swapping by considering the nodes from the neighborhood vehicle routes. In the r-opt technique, the total of r number of arcs is exchanged between the depots. The larger the value of r, the better the solution becomes. But the complexity increases as the operations required to test all the possible exchanges of r-arcs is proportional to n^r, where n is the number of nodes to be visited. This limits r to 2 or 3 which is commonly used.

There are two fundamental ways to deal with the variants of the TSP/MTSP/VRP. The first is the expansion of the MTSP into the standard TSP and the second is the simplification of the problem by using the approach of the cluster-first-route-second. Both the approaches are discussed below.

Bellmore and Hong [29] as well as Wacholder et al. [30] were the earlier ones to apply ANN for solving the MTSP by converting it into the standard TSP through the introduction of $m - 1$ new imaginary/dummy bases/depots. This makes the total number of nodes to be $n + m - 1$. This approach is further analyzed in [29, 31–34] and indicated that it worsens the situation by increasing the size of the problem, to almost double. The extended cost matrix contains at least $m - 1$ identical columns and rows. The problem becomes computationally tedious as compared to the standard TSP with the same number of nodes, i.e. $(n + m - 1)$. Bellmore and Hong [29] as well as Wacholder et al. [30] provided important observation that when used with ANN this approach becomes computationally very expensive and could converge to only a valid and not an optimal solution.

Other similarly transformed TSP has been attempted by a few researchers using the ACO [32] and the GA [11]. The results demonstrated in both are encouraging and comparable but the performance declines as the number of nodes increase. This is because the problem size and the associated computational load increase with the addition of nodes. Zhu et al. [35] attempted to solve the Single depot VRPTW using the PSO by such TSP transformation. In this approach, the transformed standard TSP which is a route (sequence of nodes) of the vehicle is represented as a particle of $n + m - 1$ dimensional vector. The formation of the particles is followed by the normalization in order to convert the particle sequences into integers or index sequences to compute the route cost.

An approach that reduces the computational load is the decomposition approach referred to as cluster-first-route-second [12, 18, 36–39]. The formation of the clusters of the nodes is based on the distances between them, the problem specific constraints such as turning radius limitations of the UAVs, [37] etc. It is followed by the formation of routes in the individual clusters and further modified for local improvement [36]. The alternative heuristic can be route-first-cluster-second where the longer routes can be constructed first followed by the shorter routes forming the clusters [19]. Gambardella et al. [40] exploited this approach to reduce the computational load using two Ant Colony Systems (ACSs) solving the VRPTW. The first ACS is for the minimization of the number of vehicles covering the maximum numbers of nodes per vehicle, i.e. one cluster is formed per vehicle. Once a feasible solution with this ACS is formed, a second ACS is used to minimize the traveling time of the vehicles covering all the points in the respective clusters, keeping the number of vehicles fixed. Both the colonies use the standard ACS framework.

Similarly, Reimann and Ulrich [38] presented a decomposition approach, in which the savings principle is used to form a complete solution with clusters of nodes. It is followed by the Sweep algorithm which exploits the information about the spatial coordinates and the centre of gravity of the predefined clusters. The swapping of nodes, 2-opt techniques are also used for the local improvement. The results of this decomposition approach reveal that it can accommodate larger size problems as the clustering and local improvement technique reduce the computational efforts.

Such simplification techniques become important when computationally expensive and time consuming algorithms such as GA is used for solving combinatorial optimization problems such as the MTSP/VRP. The key part in the GA is the formation of the chromosomes population which plays an important role in the efficiency of the algorithm. Hence the search does not start from a single solution but from a population of solutions. Furthermore, with n number of nodes, the possible solutions or routes joining the nodes are as high as $n!$. Generating such possible solutions is quite expensive if done using the permutations, making the approach quite tedious at the first instance. This makes the GA limited to a small number of nodes and to a limited number of applications, such as the minimization of the newspaper printing time [10], where the number of nodes is quite few. On the other hand, if these n numbers of nodes are divided into k clusters, then the possible number of permutations required to form the chromosomes reduces to $(n/k)!\ k$ which is comparatively fewer than the approach without clusters. A few other approaches [41, 42] followed the approach of clustering nodes around the depots based on the Euclidean distance between them.

The type of ANN referred to as the Self-Organizing Map (SOM) is used to cluster/categorize/group the unorganized data. This makes the SOM useful for both the MTSP [31] and the VRP. This work is associated with the local optimization technique such as the 2-opt updating the routes by only considering the nodes in the neighborhood zones/clusters. These clusters need to be very compact in order to reduce the computational overhead. This is a disadvantage of this approach restricting it to a fewer number of cities and also reducing the convergence speed. It outperforms the elastic net approach proposed by Vakhutinsky and Golden [43] in terms of computational time because the elastic net approach considers all the nodes while updating the routes representing heavy interconnection.

This section described the various local improvement techniques necessary to be used along with the optimization algorithms. The next section summarizes these algorithms used to solve the MTSP/VRP.

3.1.3.2 Various Algorithms for Solving the MTSP

In the general ACO technique presented by Bullnheimer et al. [44], the complexity in coding the ant trails and the associated local and the global updating makes the convergence quite slow and uncertain. This becomes worse when used NP-hard problems like the MTSP/VRP and needs to be used along with the heuristic techniques such as swapping of nodes, r-opt technique, node insertion, candidate list, etc. For example, Reimann and Ulrich [38] demonstrated the application of ACO for solving the VRPBTW by incorporating a specially developed insertion technique for accommodating the linehaul and backhaul constraint. It is followed by local swapping. For a short discussion on some of the insertion and local swapping approaches as well as the one used in the current work, refer to Sects. 3.2.3 and 3.2.5.

As the PSO is weak in local search, it may get trapped in local minima. Special hybridization techniques are needed to jump out of the local optima [28]. Generally, the parameter of the PSO referred to as 'inertia weight' is used to jump out of the local optima. But this becomes quite ineffective when used for solving combinatorial optimization problems such as MTSP/VRP [35] and suggested an alternative to increase the number of particles to at least 10 times more than the number of nodes, resulting in computational overload. In addition, increasing the number of iterations does not improve the solution when the number of particles reaches above a certain level. Furthermore, in the applications of the VRP, the handling of the capacity constraints increases the computational overload further and also needs swapping, insertion, etc. techniques as repair work to restore feasibility [28].

Zhu et al. [35] in their recent work on PSO directly incorporated the penalty coefficient into the objective function to move the solution towards feasibility. Chen et al. [28] presented a hybridization approach by combining the Discrete PSO (DPSO) with a local improvement/search technique such as SA. The job of the SA is to swap the nodes from the neighborhood vehicle routes. Similar to the mutation operation in GA, SA exchanges pairs in a vehicle route. The results indicate that the hybrid DPSO outperforms the GA used with the 2-opt technique.

According to Tang et al. [11], the GA also needs special attention in using the crossover operators solving the MTSP/VRP. This is because the direct exchanging of the parts of the parent solutions may produce infeasible solutions (a node may appear more than once or may not appear at all). This necessitates repair work to be done to make the offspring solutions feasible. The worse part is that the repair work may destroy the characteristics of the parent solutions as well as the offspring solutions. Eventually the algorithm becomes very slow due to the delayed convergence. In order to balance between the rapid convergence and avoidance of entrapment in local optima due to premature convergence, the general GA is modified and referred to as the Modified GA (MGA). In the MGA, the best chromosome or solution found so far is accepted as one of the parent solutions and another parent is selected randomly. The crossover between these two parents is done using a 'cycle crossover' (CX) method with a high possibility of retaining the characteristics of the best parent [11].

The performance of the steady state GA, in which the best found offspring replaces the worst chromosome in the population, may become worse when used with combinatorial optimization problems like the MTSP/VRP. In order to make this approach faster at the first instance, the chromosomes are formed by inserting the nodes to form the vehicle routes [45]. The crossover and mutation may produce infeasible solutions. Similar to Zhu et al. [35], Goldberg [45] demonstrated the complete avoidance of the repair work is totally by incorporating the penalty coefficient into the objective function. The results using this modified steady state GA indicates that a large population and high computational efforts are needed to reach convergence.

The computationally exhaustive approach such as ANN is first time used by Hopfield et al. [46] solving the TSP. A strong analysis is provided in [30, 47] which highlighted that the ANN approach is quite complex and unable to produce guaranteed MTSP solution. It is also not good at handling the capacity constraints and time window constraints and is essentially limited to spatial applications [19].

Durbin and Willshaw [3] proposed a unique type of SOM referred to as the elastic net approach solving the standard TSP which is also used to solve the MTSP [43]. The approach works on the basis of the pulling and attracting forces acting on the rubber band. The approach is essentially limited to the spatial applications minimizing the traveling distance between various points on the plane.

The tabu search is essentially a local search technique, improving the available feasible solution using the memory of the points visited previously. This avoids the repeated visiting of points for the predefined upcoming steps. Local search techniques such as diversification and intensification are used which are basically the combinations of the exchange of arcs in a depot and within the depots [20]. The tabu search approach finds the optimum when used with the limited class of problems such as MDVRP because of its tendency of getting trapped in local minima [26, 48–50]. This is because, although the memory of the visited points makes the algorithm efficient, some points which would have been useful at some moment may not be available. The next section discusses the application of the MTSP/VRP to the path planning of the MUAVs.

3.1.3.3 Solving Multiple Unmanned Aerial Vehicles (MUAVs) Path Planning Problem as a MTSP/VRP

The path planning/routing of MUAVs is an important potential application but there has been very few published work in the area of applying the MTSP/VRP to the MUAVs. The possible application areas are planning the path of MUAVs to visit the target points the locations of which are known in advance, planning the path for surveillance [39], etc. In addition, Klesh et al. [51] highlighted important military applications such as security intelligence, reconnaissance and surveillance with the objective of the total travel time and distance minimization of the multiple UAVs. This underscores the suitability of the MTSP approach for the above objective. In the practical sense, the implementations by Shima et al. [36], Rathinam et al. [37]

and Yadlapalli et al. [52] considers the famous Dubin's car model [53] in order to take into account the dynamic constraints such as the turning radius of the vehicles.

The fundamental approaches discussed previously in Sects. 3.1.3.1 and 3.1.3.2 are also found in the published work on MUAVs. The UAV path planning to visit the predefined target locations [37] is based on the spanning tree formation using the Prim's algorithm, which is basically an insertion algorithm iteratively adding the nodes. Shima et al. [36] combined the cluster-first-route-second and insertion technique in which the clusters belonging to the vehicles are formed by iteratively adding unselected nodes which further avoided the repair work. The best possible routes in the individual clusters optimized using the Concorde TSP Solver [54]. The computationally exhaustive tool such as Branch and Bound (B & B) technique with best-first-search approach is used. Also the memory requirement for the B & B technique is very high and may worsen if number of nodes increase. Depending on the fuel capacity, every Unmanned Combat Aerial Vehicle (UCAV) in a fleet is assigned a cluster of target nodes using the tabu search technique [39]. Christofides et al. [33] proposed the approach of cluster-first-route-second, in which once the clusters are finalized the corresponding routes are formed using the B & B approach and further combined with a specially developed insertion heuristic technique. The approach was quicker to converge but produced sub-optimal solution.

The above discussion on the various techniques and algorithms used so far in solving the MTSP/VRP indicates that they become slow as the problem size increases and limits the problem size. On the other hand, the distributed, decentralized approach can divide the computational load which can be handled more effectively and moreover may accommodate larger size problems. As a part of this research effort, PC is applied to the Multi Depot MTSP (MDMTSP) [1] in order to test its effectiveness in this area. The next few sections discuss the implementation of constrained PC solving the two test cases of the MDMTSP.

3.2 Solution to the Multiple Depot MTSP (MDMTSP) Using PC

In the proposed Multi Depot MTSP (MDMTSP), the traveling vehicles were considered as autonomous agents. These vehicles were assigned the routes, where every route was considered as a strategy. These routes were formed from randomly sampled nodes. This is in contrast to the Rosenbrock function problem in Sect. 2.5, where every agent randomly selected strategies within the range of values specified to each agent.

The sampling of routes was an important step in applying PC to the MTSP. The sampling procedure is explained in the following section by an illustrated example with 3 vehicles and 15 nodes.

Table 3.1 Initial sampling using a 3 × 7 grid

Vehicle 1		3	5		15		1
Vehicle 2	4	8	11	10		13	
Vehicle 3	6	9		2	12	14	7

3.2.1 Sampling

In problems like the MTSP, there is a strong possibility that some nodes may get sampled by more than one vehicle and also some may not get selected. This was avoided in the first sampling phase forming the individual agent routes. As there were 3 vehicles and 15 nodes, a grid of 3 × 7 was formed making sure that every vehicle is assigned at least one node. This is shown in Table 3.1 and the procedure is explained below.

The first node (node 1) randomly selects a cell from the grid, thus assigning itself to the corresponding vehicle. The cell once selected becomes unavailable for further selection by the remaining nodes. This process continues until all the nodes are assigned to the vehicles. The nodes assigned to a particular vehicle form an individual vehicle route. These routes are the strategies randomly assigned to the vehicles. According to the grid shown in Table 3.1, 4 nodes are assigned to vehicle 1 forming a route/strategy represented as D1-3-5-15-1-D1. Similarly, route/strategy formed by vehicle 2 is D2-4-8-11-10-13-D2 and by vehicle 3 is D3-6-9-2-12-14-7-D3, where D1, D2 and D3 represent the depots corresponding to the vehicles.

The above process continued for 45 times forming 45 possible combinations covering all the nodes using three vehicles. Out of these 45 combinations, the first 15 were assigned to vehicle 1 forming 15 combined strategy sets. Similarly, the next 15 combined strategy sets were assigned to vehicle 2 and the remaining 15 to vehicle 3. An example of a combined strategy set for a vehicle is represented as

$$\mathbf{Y} = \{[3 \quad 5 \quad 15 \quad 1], [4 \quad 8 \quad 11 \quad 10 \quad 13], [6 \quad 9 \quad 2 \quad 12 \quad 14 \quad 7]\} \tag{3.1}$$

According to the PC framework, in the combined strategy set shown in Eq. (3.1), the first route D1-3-5-15-1-D1 was assumed to be selected by vehicle 1 while the remaining two routes were guessed by it as the strategy sets selected by the remaining two vehicles. In order to reduce the computational load and reach convergence faster, the number of combined strategies/routes per vehicle can be limited to a small number (15 in the current example). This makes it clear that there were 15 strategies/routes considered per vehicle and they were assigned uniform starting probability values (1/15 to each route in the current example) by the corresponding vehicle.

3.2.2 Formation of Intermediate Combined Route Set

Solving this problem according to the algorithmic procedure explained in this chapter, every vehicle converged to a probability distribution clearly distinguishing

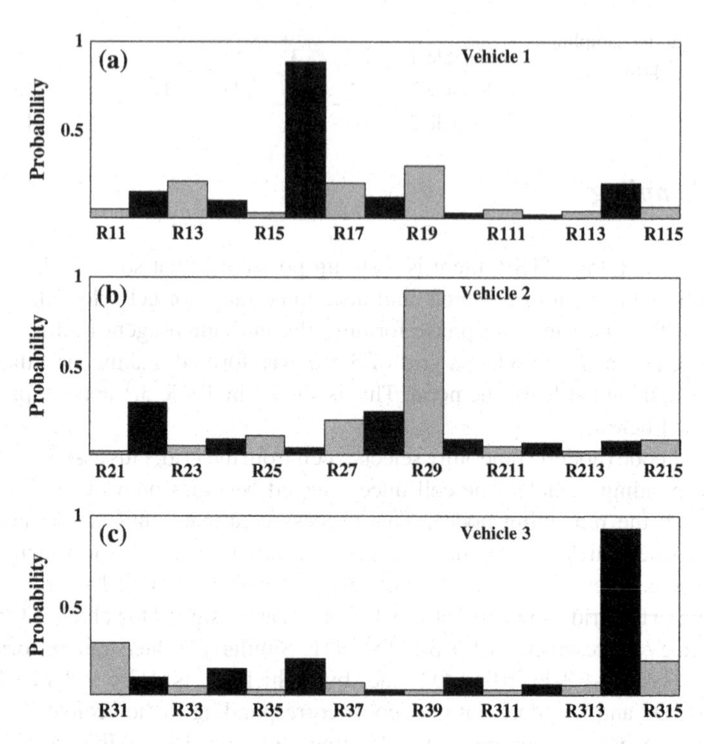

Fig. 3.1 Illustration of converged probability plots

the most favorable route i.e. the one with the highest probability value. As discussed in the preceding Sect. 3.2.1, every vehicle has a strategy set with randomly sampled routes as its strategies. These strategies are represented on the X-axis of the probability distribution plots shown in Fig. 3.1. The routes are denoted in the form R##. For example, R11 represents route 1 of vehicle 1, while R115 represents route 15 of vehicle 1. Similarly, R21 represents route 1 of vehicle 2 and so on.

As shown by an example illustration in Fig. 3.1, vehicle 1 converges to R16, vehicle 2 converges to R29 and vehicle 3 converges to R314. These most favorable routes are joined to form the 'intermediate combined route set' represented as $\{[R16], [R29], [R314]\}$.

This intermediate combined route set (also referred to as the 'intermediate solution') has a high probability of being infeasible. Infeasibility can be caused by one or more of the following four reasons:

1. The total number of nodes in the intermediate solution is less than the total number of available nodes.
2. A node has been selected by more than one vehicle, causing repetition error.
3. Some nodes are left unselected because of reason 2.

4. The total number of nodes in the combined strategy/route set formed from the most favorable strategies/routes is more than the total number of available nodes.

The above reasons arise because of the nature of the PC algorithm. The most favorable routes selected by each vehicle are generally from different combined strategy/route sets belonging to the individual vehicles. Reason 1 was avoided by repeating the sampling process until the valid intermediate solution is achieved. The next couple of sections discuss the implementation of the heuristic techniques employed to deal with reasons 2–4. Although applied to the entire intermediate solution, for simplicity and better clarity the heuristic techniques are described only for two vehicles' most favorable routes.

3.2.3 Node Insertion and Elimination Heuristic

The insertion heuristic was carried out in order to deal with reasons 2 and 3. Generally in the insertion technique a node is iteratively added to the available route and checked for feasibility and improvement. Shima et al. [36] also deployed a similar insertion technique where the Branch and Bound technique was implemented to perform Best-First Search for Unmanned Aerial Vehicle (UAV) path planning. The insertion heuristic steps 1–4 followed in this report are explained below.

1. The number of nodes of the combined strategy/route set formed from the most favorable strategy/route set was counted.
2. The missing nodes were identified.
3. The repeated nodes were replaced by the missing ones based on the total cost improvement. It is illustrated in Fig. 3.2 with an example of 2 vehicles and 11

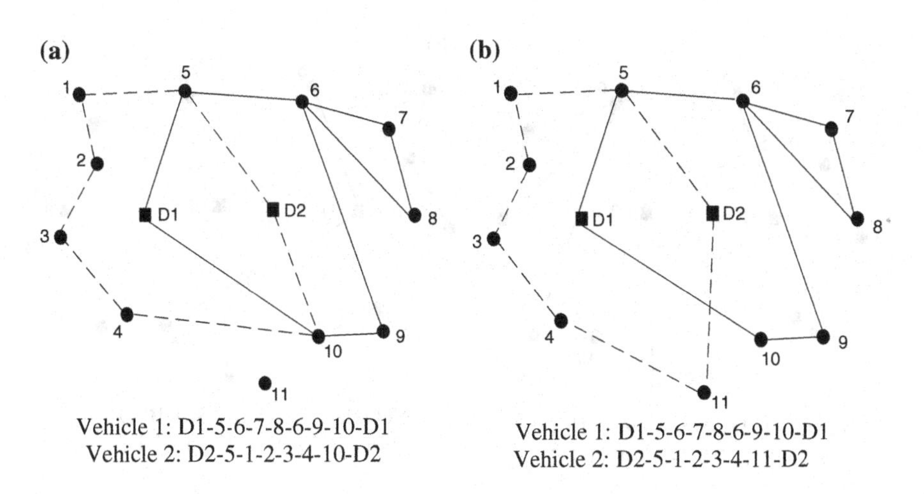

Fig. 3.2 Insertion heuristic

nodes. Figure 3.2a indicates that nodes 5 and 10 are visited by both vehicles and node 11 is left unselected. Accordingly, there are four ways to insert node 11 into a route. These are inserting it in place of the repeated node 10 or 5 along the route belonging to either vehicle. Figure 3.2b shows the lowest cost combined route/strategy set among these four different ways.

4. The steps from 1 to 3 are repeated until all the nodes are covered and appear in the intermediate solution. If the number of nodes in the intermediate solution is equal to the total number of nodes, it indicates that the solution is feasible and therefore accept the current solution and skip to step 6; otherwise continue with step 5.
5. As mentioned previously in reason 4, in some cases it is found that although all the nodes are covered, the number of nodes in the intermediate solution is more than the total number of nodes. It is an indication that some nodes are visited more than once. The elimination heuristic steps are as follows:

   5.1 The repeated nodes are identified.

   5.2 For the inter-vehicle repetition, the elimination of the repeated node from the different vehicle routes is accepted if the new solution leads to improvement. It is illustrated in Fig. 3.3a.
   Continuing from Fig. 3.2b, there are two possible remedies to inter-vehicle repetition involving node 5. The first one is assigning node 5 to the route belonging to vehicle 1 and second is assigning it to the route belonging to vehicle 2. Figure 3.3a shows the better of the two options.

   5.3 The intra-vehicle repetition was treated based on the elimination of the repeated node and the associated improvement in the feasible solution. Continuing from Fig. 3.3a, there are two possible remedies to intra-vehicle repetition to eliminate the repeated node 6. The first one is forming the route of vehicle 1 as D1-5-7-8-6-9-10-D1 and the second option is D1-5-6-7-8-9-

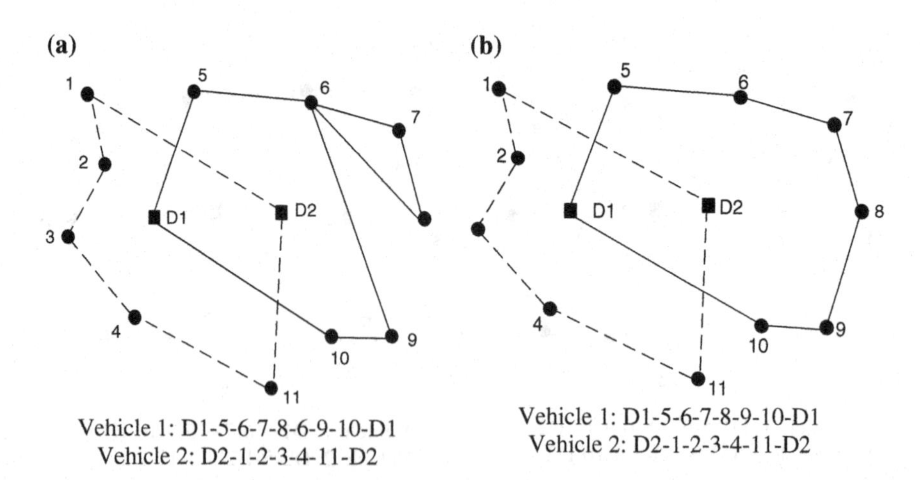

Fig. 3.3 Elimination heuristic

10-D1. The second option shown in Fig. 3.3b is the better one and valid as well. If the number of nodes covered in the intermediate solution is equal to the total number of nodes, it indicates that the solution is feasible. This intermediate solution is accepted as the current solution.

In this way, the most favorable routes were modified by repairing the infeasible intermediate solution to convert it into the feasible intermediate solution.

3.2.4 Neighboring Approach for Updating the Sample Space

The standard procedure of updating the sample space and further sample in the neighborhood of the most favorable strategy is explained in Step 7 of Sect. 2.2.1. It was applied in the MDMTSP solution procedure only when the feasible solution was achieved. The feasibility here refers to the solution having all the nodes selected with no repetition and at least one node is selected by each vehicle. The corresponding heuristic techniques are discussed in Sect. 3.2.3.

Instead of using the interval factor explained in Step 7 of Sect. 2.2.1, neighboring radii were used to update the sample space in the solution procedure of MDMTSP. For the first few iterations, sampling was carried out in the neighborhood of the nodes of a route by considering very large neighboring radii covering all the nodes. This facilitates every vehicle to select any available node and helps the algorithm to have an expanded search. The neighboring radii were reduced progressively only when there was no change in the solution for a predefined number of iterations. It is important to mention here that an identical neighboring radius was applied to every node of a route corresponding to the particular vehicle. Moreover, the nodes covered in that neighboring zone were eligible to be selected by the vehicle for further iterations. An example with arbitrary neighboring radius at some arbitrary iteration is shown in Fig. 3.4. The circles around the nodes belonging to vehicle 1 represent the neighboring zones. The nodes (5, 6, 7, 8, 9, 10 and 11) in these zones are available to be sampled randomly for vehicle 1 for the next iteration.

The procedure from Sects. 3.2.1 to 3.2.4 was repeated until there is no change in the final solution or for a predefined number of iterations i.e. until convergence. On convergence, the node swapping heuristic technique was applied. This is discussed in the next section.

3.2.5 Node Swapping Heuristic

The heuristic of swapping the set of arcs is one of the techniques used in the multi-depot case by Miliotis [55] and represented as inter-depot mutation by exchanging a set of arcs. In the current work presented here, an approach similar to that in [55]

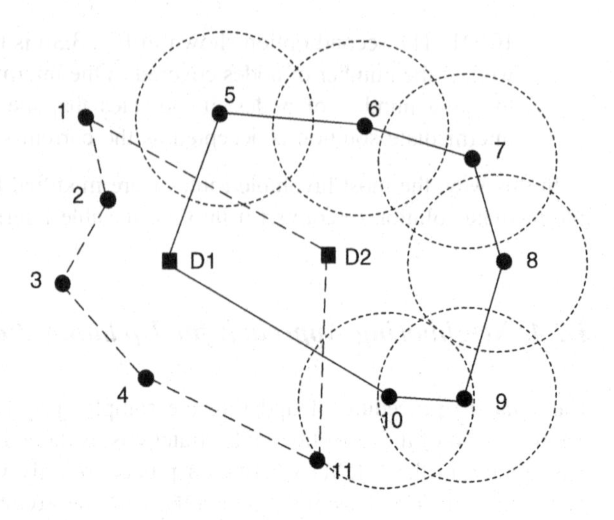

Fig. 3.4 Neighboring approach

but swapping the entire set of nodes was implemented when there was no change in the current solution for a predefined number of iterations. The entire sets of nodes belonging to two depots were exchanged and the improved solution was accepted. This heuristic proved to be very useful for the PC algorithm solving the two cases of the MDMTSP presented in the following sections. This heuristic also helped in jumping out of local optima. Continuing from the feasible solution shown in Fig. 3.3b, the node swapping is demonstrated in Fig. 3.5.

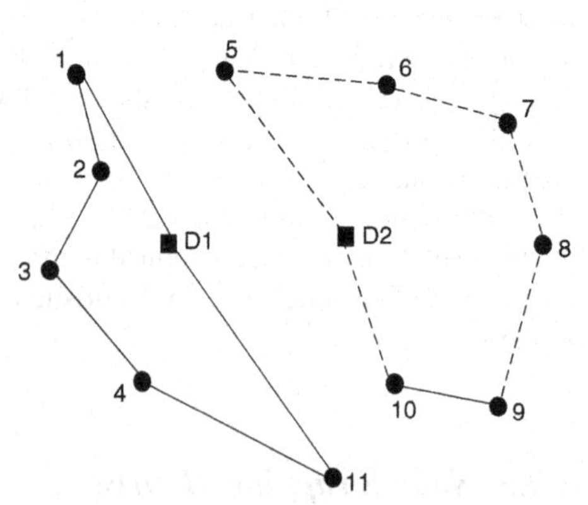

Fig. 3.5 Swapping heuristic

3.3 Test Cases of the Multiple Depot MTSP (MDMTSP)

PC accompanied with the above described heuristic techniques was applied to two test cases of the MDMTSP. Similar to the Rosenbrock function, the test cases were coded in MATLAB 7.4.0 (R2007A) on Windows platform using Pentium 4, 3 GHz processor speed and 512 MB RAM. Furthermore, for both the cases the set of parameters chosen was as follows: (a) individual vehicle sample size $m_i = 15$, (b) interval factor associated with the neighboring radius $\lambda_{down} = 0.05$.

Case 1 is presented in Fig. 3.6a where three depots are placed 120° apart from one another on the periphery of an inner circle. Fifteen equidistant nodes are placed on the outer circle. The angle subtended between adjacent nodes is 24°. The diameters of the inner circle and the outer circle are 10 and 40 units, respectively. One traveling vehicle is assigned per depot. The vehicles start their journeys from their assigned depots and return to their corresponding depots. According to the geometry of the problem and also assuming that the problem is symmetric (i.e. the cost of traveling between any two nodes is the same in both directions), the true optimum traveling cost is attained when vehicle 1 at depot 1 traveled the route D1-3-4-5-6-7-D1, vehicle 2 at depot 2 traveled the route D2-8-9-10-11-12-D2, and vehicle 3 at depot 3 traveled the route D3-13-14-15-1-2-D3. The true optimum solution was achieved by the PC approach and is plotted in Fig. 3.6b.

According to the coordinates of the nodes and the depots, the optimum cost of traveling each of these routes is 134.7870 units. Hence, the optimum total traveling cost is 404.3610 units. A total of 50 runs were conducted for Case 1 and the true optimum was reached in every run. The time to reach the optimum varied between 0.35 and 3.83 min with an average time of 2.09 min. The convergence plot of one of the runs is shown in Fig. 3.7 in which the solution at iteration 21 was accepted as

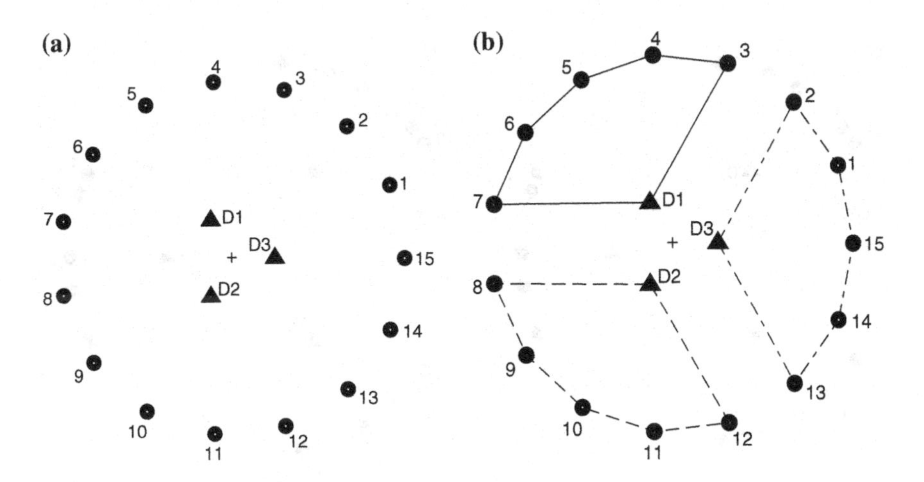

Fig. 3.6 Test case 1. **a** Case 1. **b** Solution to case 1

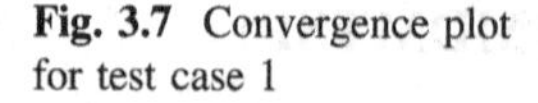

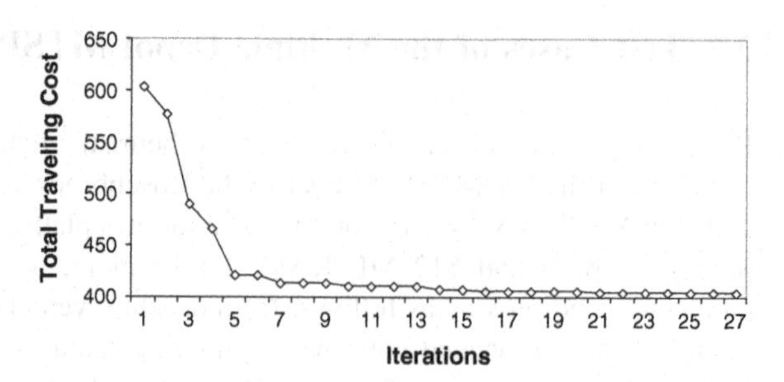

Fig. 3.7 Convergence plot for test case 1

the final solution as there was no change in the solution for a considerable number of iterations.

Case 2 is presented in Fig. 3.8a where three depots are placed 120° apart from one another on the periphery of an inner circle. Fifteen nodes are placed on the outer circle. These nodes are arranged in three clusters. In every cluster, the nodes are intentionally placed at uneven angles apart. The diameters of the inner circle and the outer circle are 5 and 10 units, respectively. Similar to Case 1, one vehicle is assigned per depot. The vehicles start their journeys from their assigned depots and return to their corresponding depots. According to the geometry of the problem and also assuming that the problem is symmetric, the optimum traveling cost is attained when vehicle 1 at depot 1 traveled the route D1-6-7-8-9-10-D1, vehicle 2 at depot 2 traveled the route D2-11-12-13-14-15-D2, and vehicle 3 at depot 3 traveled the route D3-1-2-3-4-5-D3. The true optimum solution was achieved by the PC approach and is plotted in Fig. 3.8b.

According to the coordinates of the nodes and the depots, the optimum cost of traveling each of these routes is 20.4264 units. Hence, the optimum total traveling cost is 61.2792 units. For case 2, more than 50 runs were conducted and the true

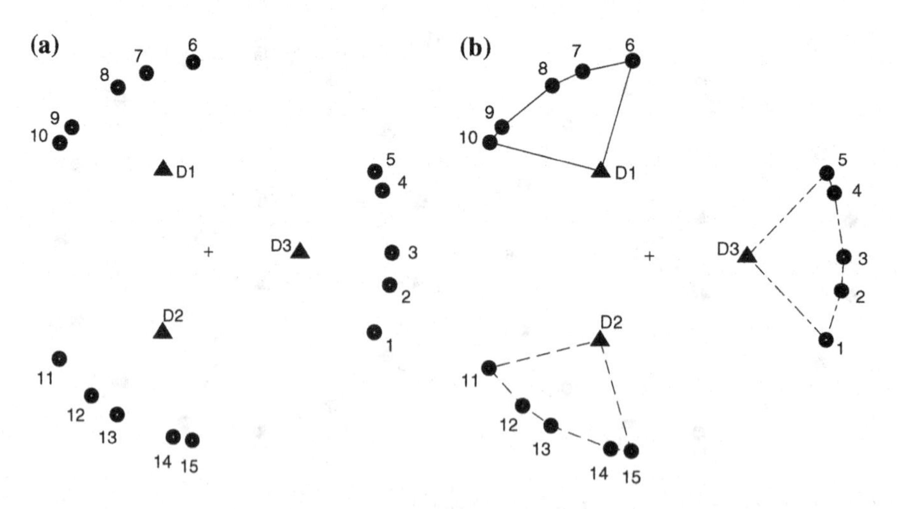

Fig. 3.8 Test case 2. **a** Case 2. **b** Solution to case 2

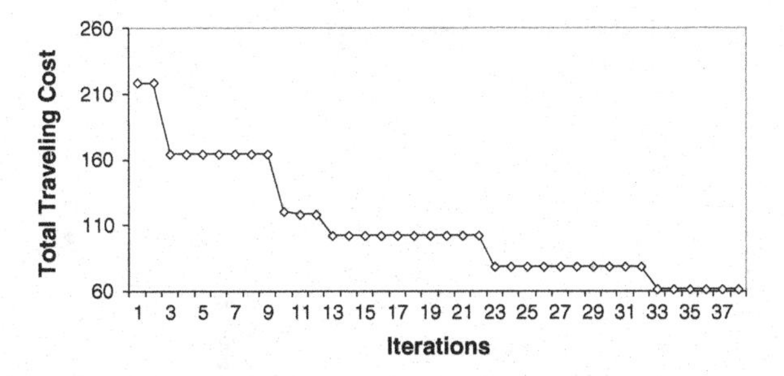

Fig. 3.9 Convergence plot for test case 2

optimum was achieved in every run. The time to reach the optimum varied between 0.76 and 3.10 min with an average time of 1.27 min. The convergence plot for one of the runs is shown in Fig. 3.9 in which the solution at iteration 33 was accepted as the final solution as there was no change in the solution for a considerable number of iterations.

3.4 Test Cases of the Single Depot MTSP (SDMTSP) with Randomly Located Nodes

Using the same procedure as the above test cases of the MDMTSP, over 30 cases of the Single Depot MTSP (SDMTSP) with 3 vehicles and 15 nodes were solved. In all the cases, the single depot and the 15 nodes were randomly located in an area of 100 × 100 units. Two sample cases are shown in Figs. 3.10a and 3.11a and the

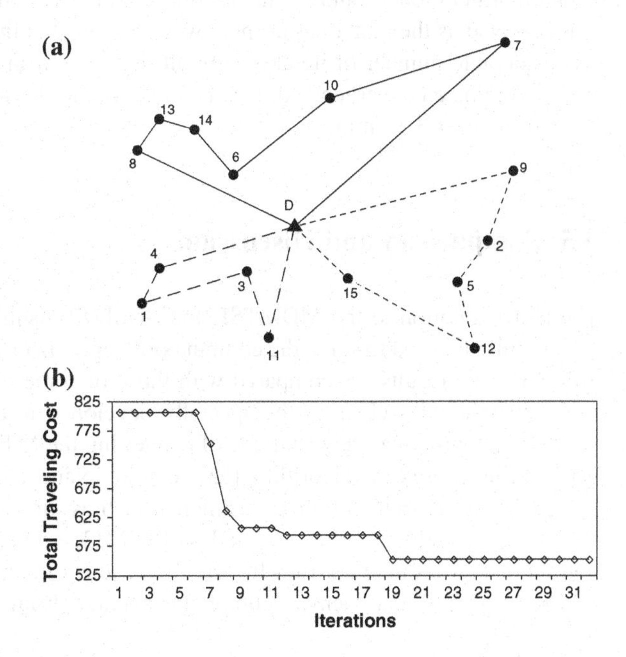

Fig. 3.10 Randomly located nodes (sample case 1)

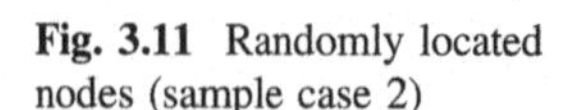

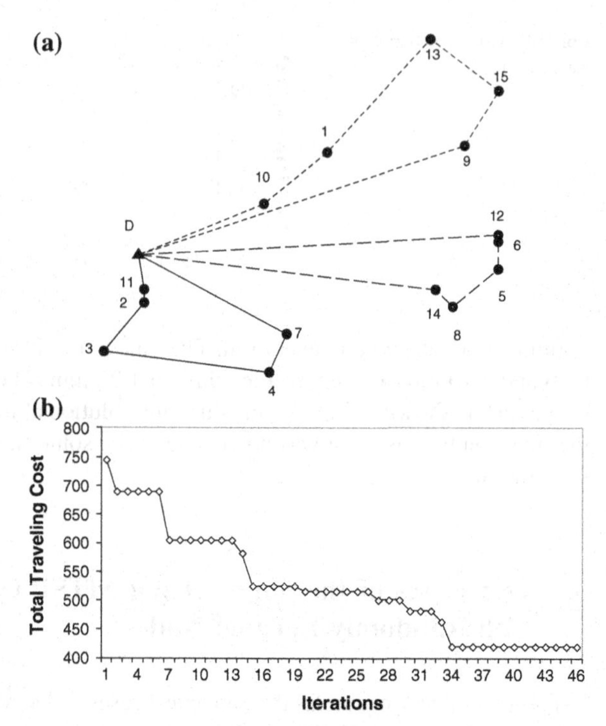

Fig. 3.11 Randomly located nodes (sample case 2)

corresponding convergence plots in Figs. 3.10b and 3.11b. In the first sample, the solution at iteration 19 and for the second sample case, the solution at iteration 33 was accepted as the final ones as there was no change in the respective solutions for a considerable number of iterations. In all the cases with randomly located nodes, the approximate convergence time was found to vary between 1 and 6 min with an average time of 3.34 min.

3.5 Comparison and Discussion

The above solution to the MDMTSP and SDMTSP using PC indicates that it can successfully be used to solve the combinatorial optimization problems. As shown in Table 3.2, the results are compared with those of some other methods solving the problem with sizes close to the specially developed test cases solved here. It is worth mentioning that the technique of converting the MTSP into the standard TSP using Miliotis reverse algorithm [55], cutting plane algorithm [34], elastic net approach [43] as well as B & B techniques such as 'branch on an arc' and 'branch on a route' with different lower bounds (LB0, LB2, etc.) in [33] did not produce the optimum solution in every run. It therefore requires the algorithms to be run for a number of trials and then accepting the average from among the trials as the

Table 3.2 Performance comparison of various algorithms solving the MTSP

Method	Nodes	Vehicles	Avg. CPU time (minutes)	% Deviation
MTSP to std. TSP [55]	20	2	2.05	–
	20	3	2.47	
	20	4	2.95	
Cutting plane [34]	20	2	1.71	–
	20	3	1.50	–
	20	4	1.44	
Elastic net[a] [43]	22	2	12.03	28.71
	22	3	13.10	74.18
	22	4	12.73	33.33
Branch on an arc with LB0 [33]	15	3	0.56	–
Branch on an arc with LB2 [33]	15	3	1.01	–
	20	4	2.32	
Branch on a route with LB0 [33]	15	3	0.44	–
PC (MDMTSP)	15 (Case 1)	3	2.09	0.00
	15 (Case 2)	3	1.27	0.00
PC (SDMTSP)	15	3	3.34	2.94

[a] Unable to reach the optimum

solution. On the other hand, every run of the PC converged to the actual optimal solution for the MDMTSP in quite reasonable CPU time. In case of the SDMTSP, the percentage deviation of total traveling cost from the average solution over 30 runs of PC was recorded to be 2.94 %. Furthermore, the CPU time varied within a very small range. The time variation exists because of the probabilistic nature of PC algorithm and random initial sampling of the routes affecting every subsequent step. As per the literature review in Sect. 3.1.3, the popular optimization techniques are computationally time consuming and expensive. It is also worth to mention that the above insertion, elimination and swapping heuristics could be easily accommodated into PC.

Besides the advantages of PC, some weaknesses associated with the above solution procedure solving the MTSP were also identified. As an effort to deal with constraints, various heuristic techniques were incorporated which served as repair techniques to push the solution into feasibility. This may not be an elegant option since they cannot be used for handling generic constraints. If complexity of the problem and related constraints increase, the repair work may become more tedious and may add further computational load. In addition, the number of function evaluations can become very large because of its dependence on the number of strategies in every agent's strategy sets. Both of these disadvantages may limit its use to smaller size problems with fewer constraints. In order to avoid this limitation and develop a more generic constraint handling technique, a penalty function approach is incorporated into PC algorithm. This is described in the following chapter.

References

1. Kulkarni, A.J., Tai, K.: Probability collectives: a multi-agent approach for solving combinatorial optimization problems. Appl. Soft. Comput. **10**(3), 759–771 (2010)
2. Kulkarni, A.J., Tai, K.: Probability collectives: a distributed optimization approach for constrained problems. In: Proceedings of IEEE World Congress on Computational Intelligence, pp. 3844–3851 (2010)
3. Durbin, R., Willshaw, D.J.: An analogue approach to the traveling salesman problem using an elastic net method. Nature **326**, 689–691 (1987)
4. Papadimitriou, C.H.: The euclidean traveling salesman problem is NP-complete. Theoret. Comput. Sci. **4**, 237–244 (1977)
5. Bektas, T.: The multiple traveling salesman problem: an overview of formulations and solution procedures. Omega **34**, 209–219 (2006)
6. Golden, B., Levy, L., Dahl, R.: Two generalizations of the traveling salesman problem. Omega **9**(4), 439–441 (1981)
7. Lawler, E., Lenstra, J., Rinnooy, K.A.H.G., Shmoys, D.: The Traveling Salesman Problem. Wiley, Chichester (1985)
8. Laporte, G.: The traveling salesman problem: an overview of exact and approximate algorithms. Eur. J. Oper. Res. **59**, 231–247 (1992)
9. Gutin, G., Punnen, A.P. (eds.): Traveling Salesman Problem and its Variations. Kluwer Academic Publishers, Dordrecht (2002)
10. Carter, A., Ragsdale, C.T.: Scheduling pre-printed newspaper advertising inserts using genetic algorithms. Omega **30**(6), 415–421 (2002)
11. Tang, L., Liu, J., Rong, A., Yang, Z.: A multiple traveling salesman problem model for hot rolling scheduling in Shangai Baoshan Iron and Steel Complex. Eur. J. Oper. Res. **124**(2), 267–282 (2000)
12. Lenstra, J.K., Rinnooy, K.A.H.G.: Some simple applications of the traveling salesman problem. Oper. Res. Quart. **26**, 717–33 (1975)
13. Lin, S.: Computer solutions of the traveling salesman problem. Bell Syst. Tech. J. **44**(10), 2245–2269 (1965)
14. Christofides, N., Eilon, S.: An algorithm for the vehicle dispatching problem. Oper. Res. Quart. **20**, 309–318 (1969)
15. Savelsbergh, M.W.P.: The general pickup and delivery problem. Trans. Sci. **29**, 17–29 (1995)
16. Svestka, J.A., Huckfeldt, V.E.: Computational experience with an M-salesman traveling salesman algorithm. Manage. Sci. **19**, 790–799 (1973)
17. Saleh, H.A., Chelouah, R.: The design of the global navigation satellite system surveying networks using genetic algorithms. Eng. Appl. Artif. Intell. **17**, 111–122 (2004)
18. Calvo, R.W., Cordone, R.: A heuristic approach to the overnight security service problem. Comput. Oper. Res. **30**, 1269–1287 (2003)
19. Potvin, J.Y.: A review of bio-inspired algorithms for vehicle routing. Stud. Comput. Intell. **161**, 1–34 (2009)
20. Renaud, J., Laporte, G., Boctor, F.F.: A tabu search heuristic for the multi-depot vehicle routing problem. Comput. Oper. Res. **23**(3), 229–235 (1996)
21. Cassidy, P.J., Bennett, H.S.: TRAMP—a multi-depot vehicle scheduling system. Oper. Res. Quart. **23**, 151–163 (1972)

22. Pooley, J.: Integrated production and distribution facility planning at ault foods. Interfaces **24**, 113–121 (1994)
23. Ball, M.O., Golden, B.L., Assad, A.A., Bodin, L.D.: Planning for truck fleet size in the presence of a common-carrier option. Decision Sci. **14**, 103–120 (1983)
24. Bell, W.J., Dalberto, L.M., Fisher, M.L., Greenfield, A.J., Jaikumar, R., Kedia, P., Mack, R. G., Prutzman, P.J.: Improving the distribution of industrial gases with an on-line computerized routing and scheduling optimizer. Interfaces **13**(6), 4–23 (1983)
25. Ghaziri, H., Osman, I.H.: Self-organizing feature maps for the vehicle routing problem with backhauls. J. Sched. **9**(2), 97–114 (2006)
26. Taillard, E.D., Laporte, G., Gendreau, M.: Vehicle routing with multiple use of vehicles. J. Oper. Res. Soc. **47**, 1065–1070 (1996)
27. Crevier, B., Cordeau, J.-F., Laporte, G.: The multi-depot vehicle routing problem with inter-depot routes. Eur. J. Oper. Res. **176**(2), 756–773 (2007)
28. Chen, A.-L., Yang, G.-K., Wu, Z.-M.: Hybrid discrete particle swarm optimization algorithm for capacitated vehicle routing problem. J. Zhejiang Univ. Sci. **7**(4), 607–614 (2006)
29. Bellmore, M., Hong, S.: Transformation of multisalesmen problem to the standard traveling salesman problem. J. Assoc. Comput. Mach. **21**, 500–504 (1974)
30. Wacholder, E., Han, J., Mann, R.C.: An extension of the hopfield-tank model for solution of the multiple traveling salesmen problem. Proc. Neural Networks **2**, 305–324 (1988)
31. Somhom, S., Modares, A., Enkawa, T.: Competition-based neural network for the multiple travelling salesmen problem with minmax objective. Comput. Oper. Res. **26**, 395–407 (1999)
32. Pan, J., Wang, D.: An ant colony optimization algorithm for multiple travelling salesman problem. In: Proceedings of First International Conference on Innovative Computing, Information and Control, pp. 210–213 (2006)
33. Christofides, N., Mingozzi, A., Toth, P.: Exact algorithms for the vehicle routing problem, based on spanning tree and shortest path relaxations. Math. Program. **20**, 255–282 (1981)
34. Laporte, G., Nobert, Y.: A cutting planes algorithms for the M-salesmen problem. Oper. Res. Quart. **31**, 1017–1023 (1980)
35. Zhu, Q., Qian, L., Li, Y., Zhu, S.: An improved particle swarm optimization algorithm for vehicle routing problem with time windows. In: Proceedings of the IEEE Congress on Evolutionary Computation, pp. 1386–1390 (2006)
36. Shima, T., Rasmussen, S., Gross, D.: Assigning micro UAVs to task tours in an urban terrain. IEEE Trans. Control Syst. Technol. **15**(4), 601–612 (2007)
37. Rathinam, S., Sengupta, R., Darbha, S.: A resource allocation algorithm for multivehicle systems with nonholonomic constraints. IEEE Trans. Autom. Sci. Eng. **4**(1), 98–104 (2007)
38. Reimann, M., Ulrich, H.: Comparing backhauling strategies in vehicle routing using ant colony optimization. CEJOR **14**(2), 105–123 (2006)
39. Shetty, V.K., Sudit, M., Nagi, R.: Priority-based assignment and routing of a fleet of unmanned combat aerial vehicles. Comput. Oper. Res. **35**, 1813–1828 (2008)
40. Gambardella, L.M., Taillard, E.D., Agazzi, G.: MACS-VRPTW: a multiple ant colony system for vehicle routing problems with time windows. In: New ideas in optimization, pp. 63–76 (1999)
41. Ombuki-Berman, B., Hanshar, F.T.: Using genetic algorithms for multi-depot vehicle routing. Stud. Comput. Intell. **161**, 77–99 (2009)
42. Thangiah, S.R., Salhi, S.: Genetic clustering: an adaptive heuristic for the multidepot vehicle routing problem. Appl. Artif. Intell. **15**, 361–383 (2001)
43. Vakhutinsky, I., A., Golden, L., B.: Solving vehicle routing problems using elastic net. In: Proceedings of IEEE International Conference on Neural Network, pp. 4535–4540 (1994)
44. Bullnheimer, B., Hartl, R.F., Strauss, C.: Applying the ant system to the vehicle routing problem. In: Metaheuristics: Advances and Trends in Local Search Paradigms for Optimization, Kluwer, Dordrecht, pp. 285–296 (1999)
45. Goldberg, D.E.: Alleles, loci and the traveling salesman problem. In: Proceedings of the First International Conference on Genetic Algorithms and Their Applications, pp. 154–159 (1985)

46. Hopfield, J.J., Tank, D.W.: Neural computation of decisions in optimization problems. Biol. Cybern. **5**, 141–152 (1985)
47. Wilson, G.V., Pawley, G.S.: On the stability of the travelling salesman problem algorithm of hopfield and tank. Biol. Cybern. **58**, 63–70 (1988)
48. Cordeau, J.F., Desaulniers, G., Desrosiers, J., Solomon, M.M., Soumis, F.: VRP with time windows. In: Toth P., Vigo D. (eds.) The Vehicle Routing Problem. SIAM, pp. 157–193 (2002)
49. Eisenstein, D.D., Iyer, A.V.: Garbage collection in Chicago: a dynamic scheduling model. Manage. Sci. **43**(7), 922–933 (1997)
50. Tung, D.V., Pinnoi, A.: Vehicle routing-scheduling for waste collection in hanoi. Eur. J. Oper. Res. **125**, 449–468 (2000)
51. Klesh, A.T., Kabamba, P.T., Girard, A.R.: Path planning for cooperative time-optimal information collection. In: Proceedings of American Control Conference, pp. 1991–1996 (2008)
52. Yadlapalli, S., Malik, W.A., Rathinam, S. Darbha, S.: A Lagrangian-based algorithm for a combinatorial motion planning problem. In: Proceedings of 46th IEEE Conference on Decision and Control, pp. 5979–5984. USA (2007)
53. Dubins, L.: On curves of minimal length with a constraint on average curvature and with prescribed initial and terminal position. Am. J. Math. **79**, 497–516 (1957)
54. Concorde, Georgia Inst. Technol., Atlanta, Software package, Ver 03.12.19 http://www.tsp.gatech.edu/concorde/index.html (2006). Accessed 23 July 2011
55. Miliotis, P.: Using cutting planes to solve the symmetric traveling salesman problem. Math. Program. **15**, 177–188 (1978)

Chapter 4
Constrained Probability Collectives with a Penalty Function Approach

There are a number of traditional constraint handling methods available, such as gradient projection method, reduced gradient method, Lagrange multiplier method, aggregate constraint method, feasible direction based method, penalty based method, etc. The penalty based methods can be referred to as generalized constraint handling methods because of their simplicity, ability to handle nonlinear constraints and can be used with most of the unconstrained optimization methods [1, 2]. They basically transform a constrained optimization problem into an unconstrained optimization problem.

A constrained PC algorithm is proposed in this chapter by incorporating the penalty function approach and associated modifications to the unconstrained PC approach described in Chap. 2. The constrained PC approach was validated by successfully solving three constrained test problems [3, 4]. The resulting constrained PC algorithm flowchart is presented in Fig. 4.1.

4.1 Penalty Function Approach

Consider a general constrained problem (in the minimization sense) as follows:

$$\begin{aligned} &\text{Minimize} \quad G \\ \text{Subject to} \quad &g_j \leq 0, \quad j = 1, 2, \ldots, s \\ &h_j = 0, \quad j = 1, 2, \ldots, t \end{aligned} \tag{4.1}$$

For each of its combined strategy set $\mathbf{Y}_i^{[r]}$, each agent i computes the corresponding system objective $G\left(\mathbf{Y}_i^{[r]}\right)$, i.e. each agent i computes m_i values of the system objective $G\left(\mathbf{Y}_i^{[r]}\right)$. In addition, each agent i computes m_i vectors of the corresponding constraints $\mathbf{C}\left(\mathbf{Y}_i^{[r]}\right)$ as follows:

© Springer International Publishing Switzerland 2015

A.J. Kulkarni et al., *Probability Collectives*, Intelligent Systems Reference Library 86, DOI 10.1007/978-3-319-16000-9_4

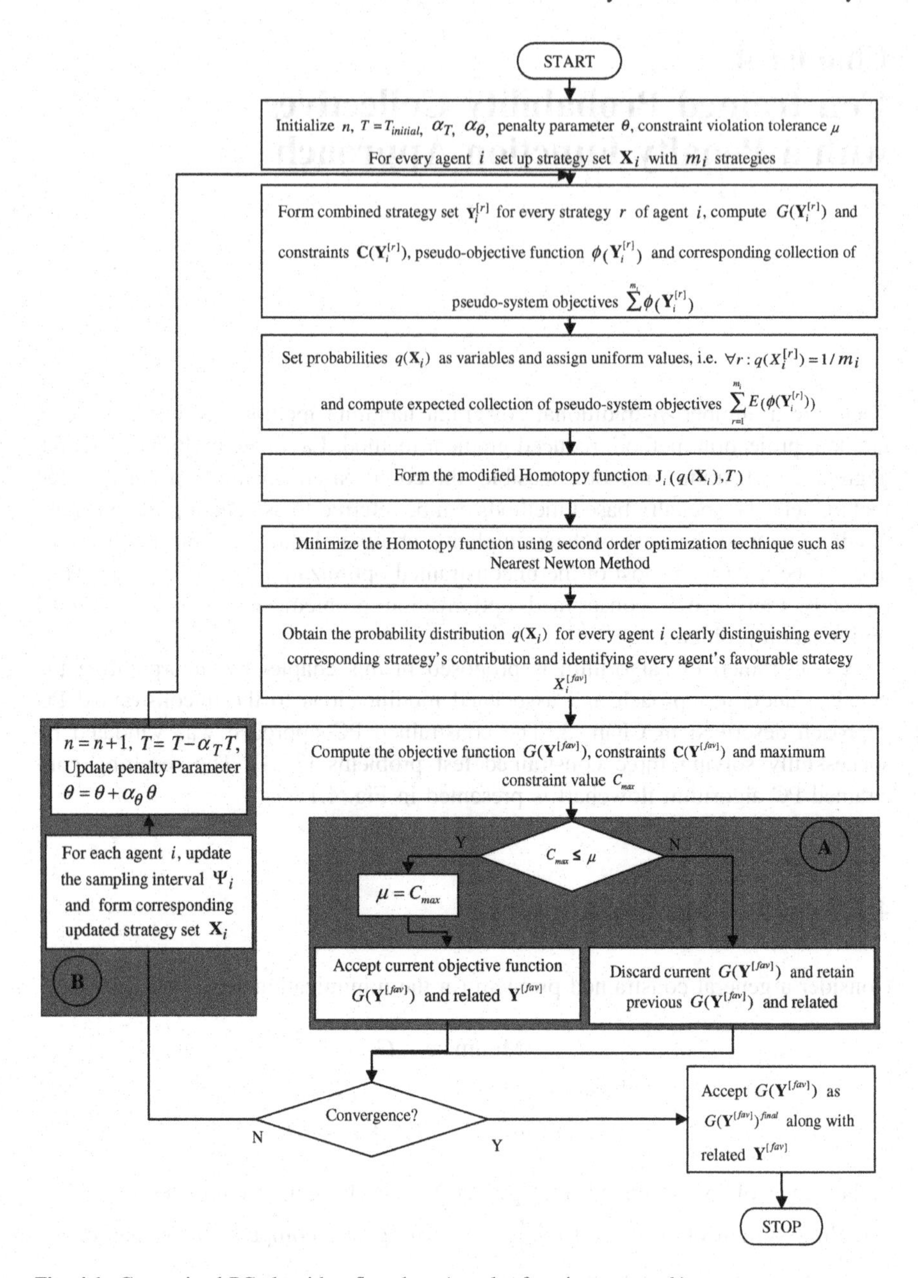

Fig. 4.1 Constrained PC algorithm flowchart (penalty function approach)

$$\begin{aligned}
\mathbf{C}\left(\mathbf{Y}_i^{[1]}\right) &= \left[g_1\left(\mathbf{Y}_i^{[1]}\right) \quad g_2\left(\mathbf{Y}_i^{[1]}\right) \quad \cdots \quad g_s\left(\mathbf{Y}_i^{[1]}\right) \quad h_1\left(\mathbf{Y}_i^{[1]}\right) \quad h_2\left(\mathbf{Y}_i^{[1]}\right) \quad \cdots \quad h_t\left(\mathbf{Y}_i^{[1]}\right)\right] \\
&\vdots \\
\mathbf{C}\left(\mathbf{Y}_i^{[r]}\right) &= \left[g_1\left(\mathbf{Y}_i^{[r]}\right) \quad g_2\left(\mathbf{Y}_i^{[r]}\right) \quad \cdots \quad g_s\left(\mathbf{Y}_i^{[r]}\right) \quad h_1\left(\mathbf{Y}_i^{[r]}\right) \quad h_2\left(\mathbf{Y}_i^{[r]}\right) \quad \cdots \quad h_t\left(\mathbf{Y}_i^{[r]}\right)\right] \\
&\vdots \\
\mathbf{C}\left(\mathbf{Y}_i^{[m_i]}\right) &= \left[g_1\left(\mathbf{Y}_i^{[m_i]}\right) \quad g_2\left(\mathbf{Y}_i^{[m_i]}\right) \quad \cdots \quad g_s\left(\mathbf{Y}_i^{[m_i]}\right) \quad h_1\left(\mathbf{Y}_i^{[m_i]}\right) \quad h_2\left(\mathbf{Y}_i^{[m_i]}\right) \quad \cdots \quad h_t\left(\mathbf{Y}_i^{[m_i]}\right)\right]
\end{aligned} \tag{4.2}$$

where s is the number of inequality constraints and t is the number of equality constraints.

In order to incorporate the constraints into the problem, each agent i forms the pseudo-objective function as follows [1, 2]:

$$\phi\left(\mathbf{Y}_i^{[r]}\right) = G\left(\mathbf{Y}_i^{[r]}\right) + \theta\left\{\sum_{j=1}^{s}\left[g_j^+\left(\mathbf{Y}_i^{[r]}\right)\right]^2 + \sum_{j=1}^{t}\left[h_j\left(\mathbf{Y}_i^{[r]}\right)\right]^2\right\} \tag{4.3}$$

where $g_j^+\left(\mathbf{Y}_i^{[r]}\right) = \max\left(0, g_j\left(\mathbf{Y}_i^{[r]}\right)\right)$ and θ is the scalar penalty parameter.

The ultimate goal of every agent i is to identify its own strategy value which contributes the most towards the minimization of the sum of these system objective values, i.e. $\sum_{r=1}^{m_i} \phi(\mathbf{Y}_i^{[r]})$, referred to as the collection of pseudo-system objectives.

In addition to the parameters listed in the unconstrained PC algorithm in Sect. 2.2.1, four more parameters are required in the constrained algorithm, namely the number of test iterations n_{test} which decides the iteration number from which the convergence condition needs to be checked, the scalar penalty parameter θ, the associated update factor α_θ for the parameter θ and a constraint violation tolerance μ. Similar to the other parameters, the values of these parameters are chosen based on the preliminary trials of the algorithm. Furthermore, at the beginning of the algorithm, the tolerance μ is initialized to a very large prescribed value. The value of μ is tightened iteratively as the algorithm progresses.

The system objective $G(\mathbf{Y}^{[fav]})$ and corresponding $\mathbf{Y}^{[fav]}$ are accepted if and only if the largest absolute component from the constraint vector $\mathbf{C}\left(\mathbf{Y}^{[fav]}\right)$ referred to as $C_{\max}$ is not worse than the constraint violation tolerance μ, i.e., $C_{\max} \le \mu$, and the value of μ is updated to $C_{\max}$, i.e., $\mu = C_{\max}$. In this way the algorithm progresses by iteratively tightening the constraint violation.

The detailed illustration of the above mentioned modifications to the unconstrained PC algorithm (discussed in Chap. 2) corresponding to the decision of acceptance/rejection of the solution, the updating of the parameters are represented in windows A and B respectively, of the constrained PC algorithm flowchart shown in Fig. 4.1.

4.2 Solutions to Constrained Test Problems

The constrained PC approach was successfully tested for solving three problems with equality and inequality constraints. These problems have already been well studied in the literature and used to compare the performance of various optimization algorithms. The various associated constraint handling techniques are discussed below in brief.

Some of the approaches [5–9] focused on overcoming the limitations of the classic penalty approach. A self adaptive penalty approach [5] and dynamic penalty scheme [7] produced premature convergence and similar to stratum approach [8] were quite sensitive to the additional set of parameters. A stochastic bubble sort algorithm in [6] attempted the direct and explicit balance between the objective function and the penalty function. Although a variety of constrained problems could be solved, it was sensitive to the associated probability parameter and needed several preliminary trials. The approach of penalty parameter was avoided using a feasibility-based rule in [9]. It required an additional fitness function and failed to produce optimum solution in every run. In addition, the quality of the solution was governed by the population in hand. The need of additional fitness function was avoided in Hybrid PSO [10] by incorporating the feasibility-based rules [9] into PSO.

The GEnetic algorithm for Numerical Optimization of COnstrained Problems (GENOCOP) specifically for linear constraints [11, 12], the PSO [13] and the Homomorphous Mapping [14] needed initial feasible solutions along with the problem dependent parameters to be set. Furthermore, for some of the problems, it was almost impossible to generate initial feasible solutions, which might need some additional techniques. The lack of diversity of the population was noticed in the cultural algorithm [15] and Cultural Differential Evolution (CDE) [16] which divide the search space into feasible, semi-feasible and infeasible regions.

An interactive approach of constraint correction algorithm [1] needed an additional constraint and several preliminary trials. The gradient repair method [17] was embedded into PSO [18] and its performance was governed by the number of solutions undergoing repair [19].

The multi-objective approach in Filter SA (FSA) [20] required numerous problem dependent parameters to be tuned along with the associated preliminary trials. Inspired from [20], GA was applied to find the non-dominated solution vector in [21]. The multi-criteria approach problem in [22–24] could not solve the fundamental problem of balancing the objective function and the constraint violations. The results of the two multi-objective approaches in [25] highlighted the necessity of an additional search bias technique to locate the feasible solutions.

This constrained PC algorithm was coded in MATLAB 7.4.0 (R2007A) and the simulations were run on a Windows platform using a Pentium 4, 3 GHz processor speed and 512 MB memory capacity. The results and the comparison with other algorithms are discussed below.

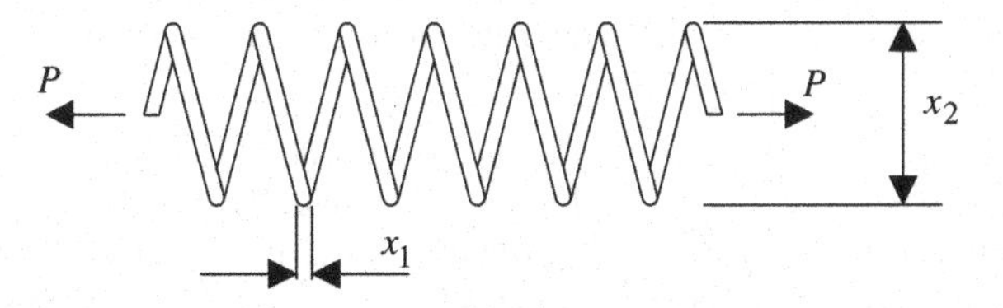

Fig. 4.2 Tension/compression spring

4.2.1 Test Problem 1

This is a tension/compression spring design problem described in [1, 5, 10, 15, 21]. It has the aim of minimizing the weight $f(\mathbf{X})$ of a tension/compression spring (as shown in Fig. 4.2) subject to constraints on minimum deflection, shear stress, surge frequency, limits on outside diameter and on design variables. The design variables are the mean coil diameter, the wire diameter x_1 and the number of active coils x_3. This problem is categorized as a quadratic problem with three variables, six boundary conditions, one linear inequality and three nonlinear inequality constraints.

The mathematical formulation of this problem can be described as follows:

$$\text{Minimize} \quad f(\mathbf{X}) = (x_3 + 2)x_2x_1^2$$

$$\text{Subject to } g_1(\mathbf{X}) = 1 - \frac{x_2^3x_3}{71785x_1^4} \leq 0$$

$$g_2(\mathbf{X}) = \frac{4x_2^2 - x_1x_2}{12566\left(x_2x_1^3 - x_1^4\right)} + \frac{1}{5108x_1^2} - 1 \leq 0$$

$$g_3(\mathbf{X}) = 1 - \frac{140.45x_1}{x_2^2x_3} \leq 0$$

$$g_4(\mathbf{X}) = \frac{x_1 + x_2}{1.5} - 1 \leq 0$$

where $0.05 \leq x_1 \leq 2,\ 0.25 \leq x_2 \leq 1.3,\ 2 \leq x_3 \leq 15$.

This problem was solved in [1, 5, 10, 15, 21]. The best solutions obtained by these approaches along with those by the PC approach are listed in Table 4.1. The PC solution to the problem produced competent results at low computational cost (computational costs of the other approaches were not available). The best, mean and worst function values found from ten runs were 0.01350, 0.02607, 0.05270, respectively. Furthermore, the average CPU time in seconds and average number of function evaluations were 24.5 and 5,214, respectively. Moreover, in every iteration, the parameters such as the number of strategies m_i and interval factor λ_{down} were 6 and 0.1, respectively. The best PC solution was within about 6.63 % of the best-reported solution [10]. The convergence plot for the best PC solution is shown in Fig. 4.3.

Table 4.1 Performance comparison of various algorithms solving problem 1

Design variables	Best solutions found					
	Coello et al. [15]	Arora [1]	Coello [5]	Coello et al. [21]	He [10]	PC
x_1	0.05000	0.05339	0.05148	0.05198	0.05170	0.05060
x_2	0.31739	0.39918	0.35166	0.36396	0.35712	0.32781
x_3	14.03179	9.18540	11.63220	10.89052	11.26508	14.05670
$g_1(\mathbf{X})$	0.00000	0.00001	−0.00330	−0.00190	−0.00000	−0.05290
$g_2(\mathbf{X})$	−0.00007	−0.00001	−0.00010	0.00040	0.00000	−0.00740
$g_3(\mathbf{X})$	−3.96796	−4.12383	−4.02630	−4.06060	−4.05460	−3.70440
$g_4(\mathbf{X})$	−0.75507	−0.69828	−0.73120	−0.72270	−0.72740	−0.74769
$f(\mathbf{X})$	0.01272	0.01273	0.01270	0.01268	0.01266	0.01350

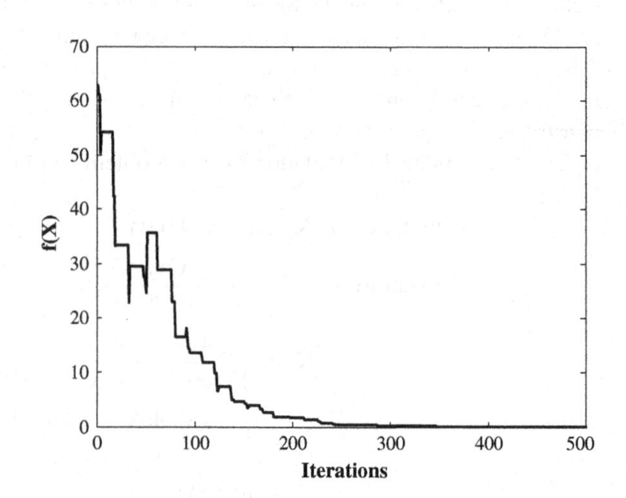

Fig. 4.3 Convergence plot for problem 1

4.2.2 Test Problem 2

This is Himmelblau's nonlinear optimization problem [26]. It has five design variables $(x_1, x_2, x_3, x_4, x_5)$ and ten boundary conditions. According to Batista et al. [27], the ratio of the size of the feasible search space to the size of the whole search space (which represent the degree of difficulty) is 52.123 %. There are six nonlinear inequality constraints out of which two are active at the optimum. The problem can be stated as follows:

Minimize

$f(\mathbf{X}) = 5.3578547x_3^2 + 0.8356891x_1x_5 + 37.293239x_1 - 40792.141$

Subject to

$g_1(\mathbf{X}) = 85.334407 + 0.0056858x_2x_5 + 0.0006262x_1x_4 - 0.0022053x_3x_5 - 92 \leq 0$

$g_2(\mathbf{X}) = -85.334407 - 0.0056858x_2x_5 - 0.0006262x_1x_4 + 0.0022053x_3x_5 \leq 0$

$g_3(\mathbf{X}) = 80.51249 + 0.0071317x_2x_5 + 0.0029955x_1x_2 + 0.0021813x_3^2 - 110 \leq 0$

$g_4(\mathbf{X}) = -80.51249 - 0.0071317x_2x_5 - 0.0029955x_1x_2 - 0.0021813x_3^2 + 90 \leq 0$

$g_5(\mathbf{X}) = 9.300961 + 0.0047026x_3x_5 + 0.0012547x_1x_3 + 0.0019085x_3x_4 - 25 \leq 0$

$g_6(\mathbf{X}) = -9.300961 - 0.0047026x_3x_5 - 0.0012547x_1x_3 - 0.0019085x_3x_4 + 20 \leq 0$

where $78 \leq x_1 \leq 102, \quad 33 \leq x_2 \leq 45, \quad 27 \leq x_i \leq 45, \quad (i = 3, 4, 5)$.

This problem was solved in [5–10, 13–20, 23, 25]. The best, mean, and worst solutions obtained by these approaches along with those from ten runs of the PC approach are listed in Table 4.2. The PC solution to this problem produced competent results at reasonable computational cost. In addition, the best PC solution was within about 0.078 % of the best-reported solution [6]. Furthermore, the average CPU time was 11 min. Moreover, in every iteration, the parameters such as the number of strategies m_i and interval factor λ_{down} were 100 and 0.01, respectively. The convergence plot for the best PC solution is shown in Fig. 4.4.

4.2.3 Test Problem 3

This problem is categorized as logarithmic nonlinear with ten variables, ten boundary conditions and three linear equality constraints. The problem can be stated as follows:

$$\text{Minimize} \quad f(\mathbf{X}) = \sum_{j=1}^{10} x_j \left(c_j + \ln \frac{x_j}{x_1 + x_2 + \cdots + x_{10}} \right)$$

$$\begin{aligned} \text{Subject to} \quad & h_1(\mathbf{X}) = x_1 + 2x_2 + 2x_3 + x_6 + x_{10} - 2 = 0 \\ & h_2(\mathbf{X}) = x_4 + 2x_5 + x_6 + x_7 - 1 = 0 \\ & h_3(\mathbf{X}) = x_3 + x_7 + x_8 + 2x_9 + x_{10} - 1 = 0 \\ & x_i \geq 0.000001, \quad i = 1, 2, \ldots, 10 \end{aligned}$$

where

$$\begin{array}{lll} c_1 = -6.089 & c_2 = -17.164 & c_3 = -34.054 \\ c_4 = -5.914 & c_5 = -24.721 & c_6 = -14.986 \\ c_7 = -24.100 & c_8 = -10.708 & c_9 = -26.662 \\ c_{10} = -22.179 & & \end{array}$$

Table 4.2 Performance comparison of various algorithms solving problem 2

Algorithm	Best	Mean	Worst	Standard deviation (SD)	Average function evaluations (FE)
Coello et al. [15]	−30665.5000	−30662.5000	−30636.2000	9.3	–
Becerra et al. [16]	−30665.5386	−30665.5386	−30665.5386	0.00000	–
Koziel et al. [14]	−30664.0000	−30655.0000	−30645.0000	–	1,400,000
Hedar et al. [20]	−30665.5380	−30665.4665	−30664.6880	0.173218	86,154
Coello [5]	−31020.8590	−30984.2400	−30792.4070	73.633	–
Chootinan et al. [17]	−30665.5386	−30665.3538	−30665.5386	0.000000	26,981
Runarsson et al. [25]	−30665.5386	−30665.3539	−30665.5386	0.000000	66,400
Runarsson et al. [6]	−30665.5390	−30665.5390	−30665.5390	1.1E-11	3,50,000
Zahara et al. [18]	−30665.5386	−30665.35386	−30665.5386	0.000000	19,658
Deb [9]	−30665.5370	–	29846.6540	–	
Farmani et al. [7]	−30665.5000	−30665.2000	−30663.3000	4.85E-01	–
Hu et al. [13]	30665.5000	30665.5000	30665.5000	–	–
Dong et al. [19]	−30664.7000	−30662.8000	−30656.1000	–	–
Ray et al. [23]	−30651.6620	−30647.1050	−30619.0470	–	35,408
He et al. [10]	−30665.5390	−30665.5390	−30665.5390	1.7E-06	
Homaifar et al. [8]	−30373.9500	–	−30175.8040	–	–
PC	−30641.5702	−30635.4157	−30626.7492	7.5455	2,780,449

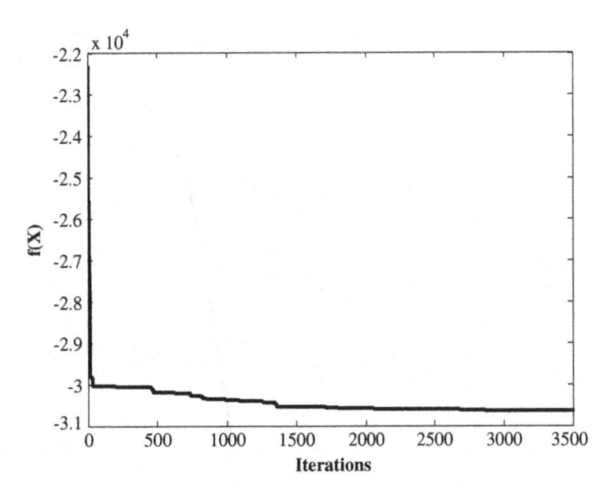

Fig. 4.4 Convergence plot for problem 2

This problem is a modified version of the chemical equilibrium problem in complex mixtures originally presented in [28]. The modified problem was solved in [12, 29]. The best solutions obtained by these approaches along with those by the PC approach are listed in Table 4.3. The PC solution to this problem produced competent results but at high computational cost (computational costs of the other approaches were not available). The best, mean and worst function values

Table 4.3 Performance comparison of various algorithms solving problem 3

Best solutions found			
Design variables	Hock et al. [29]	Michalewicz [12]	PC
x_1	0.01773548	0.04034785	0.0308207485
x_2	0.08200180	0.15386976	0.2084261218
x_3	0.88256460	0.77497089	0.6708869580
x_4	0.0007233256	0.00167479	0.0371668767
x_5	0.4907851	0.48468539	0.3510055351
x_6	0.0004335469	0.00068965	0.1302810195
x_7	0.01727298	0.02826479	0.1214712339
x_8	0.007765639	0.01849179	0.0343070642
x_9	0.01984929	0.03849563	0.0486302636
x_{10}	0.05269826	0.10128126	0.0486302636
$h_1(\mathbf{X})$	8.6900E-08	6.0000E-08	−0.0089160590
$h_2(\mathbf{X})$	0.0141	1.0000E-08	−0.0090697995
$h_3(\mathbf{X})$	5.9000E-08	−1.0000E-08	−0.0047181958
$f(\mathbf{X})$	−47.707579	−47.760765	−46.7080572120
Average function evaluations	–	–	389,546

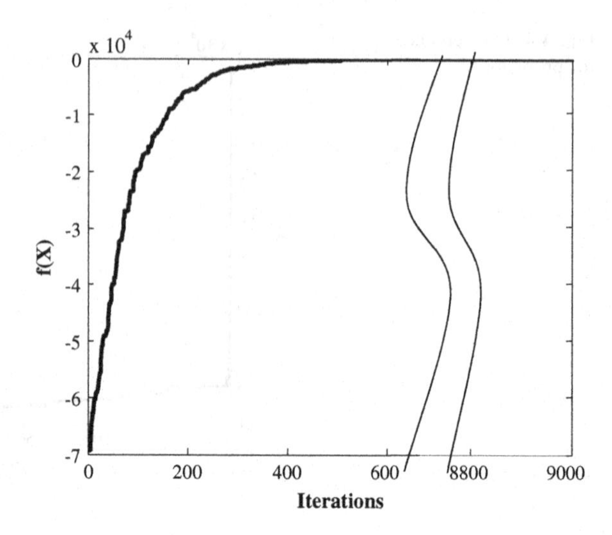

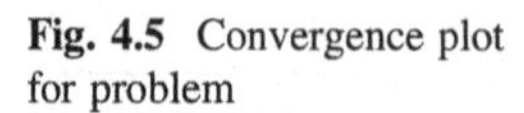
Fig. 4.5 Convergence plot for problem

found from ten runs were −46.7080572120, −45.6522267370, −44.4459333503, respectively, with standard deviation 0.7893. Furthermore, the average CPU time for PC solution was 21.60 min. Moreover, in every iteration, the parameters such as the number of strategies m_i and interval factor λ_{down} were 6 and 0.1, respectively. The best PC solution was within about 2.20 % of the best-reported solution [12]. The convergence plot for the best PC solution is shown in Fig. 4.5.

4.3 Discussion

The above solutions using PC indicated that it can successfully be used to solve a variety of constrained optimization problems. The results also demonstrated its competitiveness with those of some other methods. Furthermore, it was evident from the results that the approach is sufficiently robust and produced consistent results in every run. It implies that the rational behavior of the agents could be successfully formulated and demonstrated.

The worthy features of the above constrained PC approach are discussed here by comparing with some other methods. Approaches such as PSO [13], Homomorphous Mapping [14], GENOCOP [12] require initial feasible solution, whereas the constrained PC approach proposed here can solve the problems starting from randomly generated and completely infeasible solutions. The approach presented here can handle both equality as well as inequality constraints, whereas the approach of concurrent co-evolution in [5] is applicable only to inequality constraints with further additional computations involved. The method of GENOCOP [11, 12] is suitable only for the linear constraints while the proposed approach can handle both linear and nonlinear constraints.

The next three chapters discuss two variations of the Feasibility-based Rule as an effort to make PC a more versatile optimization approach with applications to Circle Packing Problem (CPP), discrete and mixed variable problems from the structural and mechanical engineering design domain as well as Sensor Network Coverage Problem (SNCP).

References

1. Arora, J.S.: Introduction to Optimum Design. Elsevier Academic Press, San Diego (2004)
2. Vanderplaat, G.N.: Numerical Optimization Techniques for Engineering Design. Mcgraw-Hill, New York (1984)
3. Kulkarni, A.J., Tai, K.: Solving constrained optimization problems using probability collectives and a penalty function approach. Int. J. Comput. Intell. Appl. **10**(4), 445–470 (2011)
4. Kulkarni, A.J., Tai, K.: Probability collectives: a distributed optimization approach for constrained problems. In: Proceedings of IEEE World Congress on Computational Intelligence, pp. 3844–3851 (2010)
5. Coello Coello, C.A.: Use of self-adaptive penalty approach for engineering optimization problems. Comput. Ind. **41**, 113–127 (2000)
6. Runarsson, T.P., Yao, X.: Stochastic ranking for constrained evolutionary optimization. IEEE Trans. Evol. Comput. **4**(3), 284–294 (2000)
7. Farmani, R., Wright, J.A.: Self-adaptive fitness formulation for constrained optimization. IEEE Trans. Evol. Comput. **7**(5), 445–455 (2003)
8. Homaifar, A., Qi, C.X., Lai, S.H.: Constrained optimization via genetic algorithms. Simulation **62**(4), 242–254 (1994)
9. Deb, K.: An efficient constraint handling method for genetic algorithms. Comput. Meth. Appl. Mech. Eng. **186**, 311–338 (2000)
10. He, Q., Wang, L.: A hybrid particle swarm optimization with a feasibility-based rule for constrained optimization. Appl. Math. Comput. **186**, 1407–1422 (2007)
11. Michalewicz, Z., Janikow, C.Z.: Handling constraints in genetic algorithms. Proc. Int. Conf. Genet. Algorithms **4**, 151–157 (1991)
12. Michalewicz, Z.: Numerical Optimization: Handling Linear Constraints. Handbook of Evolutionary Computation, IOP Publishing, G9.1, (1997)
13. Hu, X., Eberhart, R.: Solving constrained nonlinear optimization problems with particle swarm optimization. In: Proceedings of the 6th World Multi-conference on Systemics, Cybernetics and Informatics, (2002)
14. Koziel, S., Michalewicz, Z.: Evolutionary algorithms, homomorphous mapping, and constrained parameter optimization. Evol. Comput. **7**(1), 19–44 (1999)
15. Coello Coello, C.A., Becerra, R.L.: Efficient evolutionary optimization through the use of a cultural algorithm. Eng. Optim. **36**(2), 219–236 (2004)
16. Becerra, R.L., Coello Coello, C.A.: Cultured differential evolution for constrained optimization. Comput. Methods Appl. Mech. Eng. **195**, 4303–4322 (2006)
17. Chootinan, P., Chen, A.: Constraint handling in genetic algorithms using a gradient-based repair method. Comput. Oper. Res. **33**, 2263–2281 (2006)
18. Zahara, E., Hu, C.H.: Solving constrained optimization problems with hybrid particle swarm optimization. Eng. Optim. **40**(11), 1031–1049 (2008)
19. Dong, Y., Tang, J., Xu, B., Wang, D.: An application of swarm optimization to nonlinear programming. Comput. Math. Appl. **49**, 1655–1668 (2005)
20. Hedar, A.R., Fukushima, M.: Derivative-free simulated annealing method for constrained continuous global optimization. J. Global Optim. **35**, 521–549 (2006)

21. Coello Coello, C.A., Montes, E.M.: Constraint-handling in genetic algorithms through the use of dominance-based tournament selection. Adv. Eng. Inform. **16**, 193–203 (2002)
22. Ray, T., Tai, K., Seow, K.C.: An evolutionary algorithm for constrained optimization. In: Proceedings of the Genetic and Evolutionary Computation Conference, pp. 771–777 (2000)
23. Ray, T., Tai, K., Seow, K.C.: Multiobjective design optimization by an evolutionary algorithm. Eng. Optim. **33**(4), 399–424 (2001)
24. Tai, K., Prasad, J.: Target-matching test problem for multiobjective topology optimization using genetic algorithms. Struct. Multi. Design Optim. **34**(4), 333–345 (2007)
25. Runarsson, T.P., Yao, X.: Search biases in constrained evolutionary optimization. IEEE Trans. Syst. Man Cybern. Part C Appl. Rev. **35**(2), 233–243 (2005)
26. Himmelblau, D.M.: Applied Nonlinear Programming. Mcgraw-Hill, New York (1972)
27. Batista, B.M., Moreno Perez, J.A., Moreno Vega, J.M.: Nature-inspired decentralized cooperative metaheuristic strategies for logistic problems. In: Proceedings of European Symposium on Nature-inspired Smart Information Systems, (2006)
28. White, W.B., Johnson, S.M., Dantzig, G.B.: Chemical equilibrium in complex mixtures. Chem. Phys. **28**(5), 751–755 (1958)
29. Hock, W., Schittkowski, K.: Test examples for nonlinear programming codes. In: Lecture notes in Economics and Mathematical Systems, vol. 187. Springer, Berlin-Heidelberg-New York (1981)

Chapter 5
Constrained Probability Collectives with Feasibility Based Rule I

This chapter demonstrates further efforts to develop a generic constraint handling technique for PC in order to make it a more versatile optimization algorithm. A variation of the Feasibility-based Rule referred to as Feasibility-based Rule I originally proposed in [1] and further implemented in [2–7] were incorporated into the unconstrained PC approach from Chap. 2.

Consider a general constrained problem (in the minimization sense) as follows:

$$\begin{aligned} &\text{Minimize} \quad G \\ &\text{Subject to} \quad g_j \le 0, \quad j = 1, 2, \ldots, s \\ &\qquad\qquad\quad h_j = 0, \quad j = 1, 2, \ldots, w \end{aligned} \tag{5.1}$$

According to [8–10], the equality constraint $h_j = 0$ can be transformed into a pair of inequality constraints using a tolerance value δ as follows:

$$h_j = 0 \quad \Rightarrow \begin{cases} g_{s+j} = h_j - \delta \le 0 & j = 1, 2, \ldots, w \\ g_{s+w+j} = -\delta - h_j \le 0 \end{cases} \tag{5.2}$$

Thus, w equality constraints are replaced by $2w$ inequality constraints with the total number of constraints given by $t = s + 2w$. Then a generalized representation of the problem in Eq. (5.1) can be stated as follows:

$$\begin{aligned} &\text{Minimize} \quad G \\ &\text{Subject to} \quad g_j \le 0, \quad j = 1, 2, \ldots, t \end{aligned} \tag{5.3}$$

The Feasibility-based Rule I assisted with the perturbation approach, updating of the sampling space in the neighborhood of the favorable strategy and the modified convergence criterion are presented in Sect. 5.1 and the validation of the approach by solving two cases of the Circle Packing Problem (CPP) is presented in Sect. 5.2. In addition, attaining the true optimum solution to the CPP using PC clearly demonstrated its inherent ability to avoid the tragedy of commons [11].

© Springer International Publishing Switzerland 2015
A.J. Kulkarni et al., *Probability Collectives*, Intelligent Systems Reference Library 86, DOI 10.1007/978-3-319-16000-9_5

In addition to the two cases of the CPP, as a distributed optimization approach PC was also tested for its capability to deal with the agent failure scenario [12]. The solution highlights its strong potential to deal with the agent failure which may arise in real world complex problems including urban traffic control, formation of airplanes fleet and mid-air collision avoidance, etc.

5.1 Feasibility-Based Rule I

This rule allows the objective function and the constraint information to be considered separately. Similar to the penalty function approach presented in Chap. 4, the constraint violation tolerance is iteratively tightened in order to obtain the fitter solution and further drive the convergence towards feasibility. More specifically, at the beginning of the PC algorithm, the constraint violation tolerance μ is initialized to the number of constraints $|\mathbf{C}|$, i.e. $\mu = |\mathbf{C}|$ where $|\mathbf{C}|$ refers to the cardinality of the constraint vector $\mathbf{C} = [g_1 \quad g_2 \quad \ldots \quad g_t]$. The value of μ is tightened iteratively as the algorithm progresses.

As mentioned previously, the rule assisted with the perturbation approach, updating of the sampling space in the neighborhood of the favorable strategy and the modified convergence criterion are incorporated into the PC framework. The rule and these assisting techniques are discussed below as modifications to the corresponding steps of the unconstrained PC approach presented in Chap. 2. The following Sects. 5.1.1–5.1.3 should therefore be read in conjunction with the unconstrained PC algorithm steps described in Sect. 2.1.1.

5.1.1 Modifications to Step 5 of the Unconstrained PC Approach

The step 5 of the unconstrained PC procedure discussed in Sect. 2.2.1 is modified by using the following rule:

Any feasible solution is preferred over any infeasible solution.

Between two feasible solutions, the one with better objective is preferred.

Between two infeasible solutions, the one with fewer violated constraints is preferred.

The detailed formulation of the above rule is explained below and further presented in window A of the constrained PC algorithm flowchart in Fig. 5.1.

If the current system objective $G(\mathbf{Y}^{[fav]})$ as well as the previous solution are infeasible, accept the current system objective $G(\mathbf{Y}^{[fav]})$ and corresponding $\mathbf{Y}^{[fav]}$ as the current solution if the number of constraints violated $C_{violated}$ is less than or equal to μ, i.e. $C_{violated} \leq \mu$, and then the value of μ is updated to $C_{violated}$, i.e. $\mu = C_{violated}$.

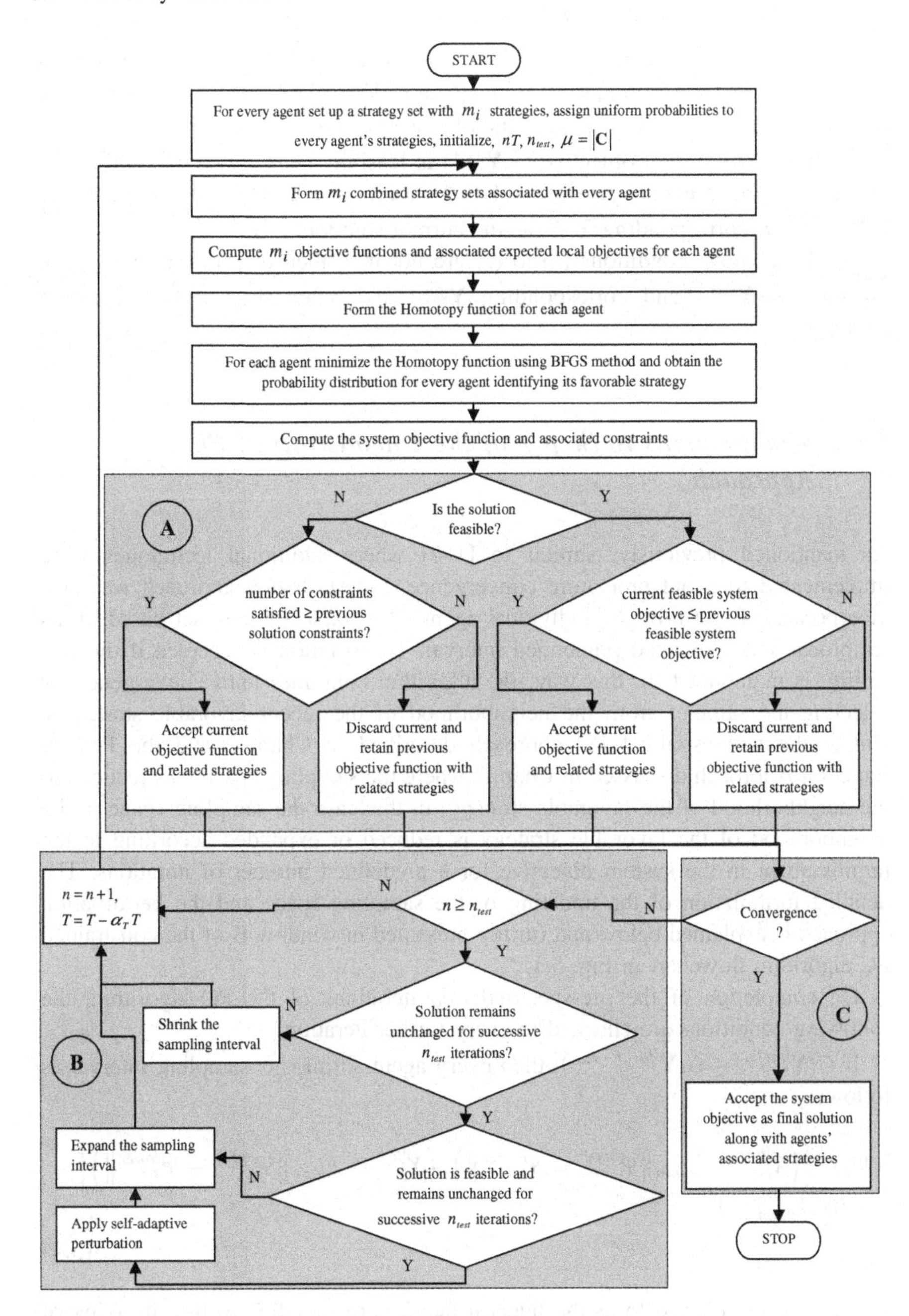

Fig. 5.1 Constrained PC algorithm flowchart (Feasibility-based Rule I)

If the current system objective $G(\mathbf{Y}^{[fav]})$ is feasible, and the previous solution is infeasible, accept the current system objective $G(\mathbf{Y}^{[fav]})$ and corresponding $\mathbf{Y}^{[fav]}$ as the current solution, and then the value of μ is updated to 0, i.e. $\mu = C_{violated} = 0$.

If the current system objective $G(\mathbf{Y}^{[fav]})$ is feasible, i.e. $C_{violated} = 0$ and is not worse than the previous feasible solution, accept the current system objective $G(\mathbf{Y}^{[fav]})$ and corresponding $\mathbf{Y}^{[fav]}$ as the current solution.

If all the above conditions (a) to (c) are not met, then discard current system objective $G(\mathbf{Y}^{[fav]})$ and corresponding $\mathbf{Y}^{[fav]}$, and retain the previous iteration solution.

5.1.2 Modifications to Step 7 of the Unconstrained PC Approach

As mentioned previously, similar to [1–7] where additional techniques were implemented to avoid premature convergence, a perturbation approach was also incorporated. It perturbs the individual agent's favorable strategy set based on its reciprocal and associated predefined interval. The solution is accepted if the feasibility is maintained. In this way, the algorithm continues until convergence by selecting the samples from the neighborhood of the recent favorable strategies. Unlike the unconstrained PC approach described in Chap. 2 and the Penalty Function Approach discussed in Chap. 4 where the sampling space was reduced to the neighborhood of the favorable strategy, in this case the sampling space in the neighborhood of the favorable strategy is reduced or expanded according to the improvement in the system objective for a predefined number of iterations. The detailed formulation of the updating of the sampling space and the perturbation approach is explained below and further presented in window B of the constrained PC algorithm flowchart in Fig. 5.1.

On completion of the pre-specified n_{test} iterations of the PC algorithm, the following conditions are checked in every further iteration.

If $G(\mathbf{Y}^{[fav],n}) \leq G(\mathbf{Y}^{[fav],n-n_{test}})$, then every agent shrinks its sampling interval as follows:

$$\Psi_i \in \left[\left(X_i^{[fav]} - \lambda_{down}\left\|\Psi_i^{upper} - \Psi_i^{lower}\right\|\right), \left(X_i^{[fav]} + \lambda_{down}\left\|\Psi_i^{upper} - \Psi_i^{lower}\right\|\right)\right],$$
$$0 < \lambda_{down} \leq 1 \tag{5.4}$$

where λ_{down} is referred to as the interval factor corresponding to the shrinking of sample space.

If $G(\mathbf{Y}^{[fav],n})$ and $G(\mathbf{Y}^{[fav],n-n_{test}})$ are feasible and $\left\|G(\mathbf{Y}^{[fav],n}) - G(\mathbf{Y}^{[fav],n-n_{test}})\right\| \leq \varepsilon$, the system objective $G(\mathbf{Y}^{[fav],n})$ can be referred to as a stable solution $G(\mathbf{Y}^{[fav],s})$ or possible local minimum.

In order to jump out of this possible local minimum, every agent i perturbs its current favorable strategy $X_i^{[fav]}$ by a perturbation factor $fact_i$ corresponding to the reciprocal of its favorable strategy $X_i^{[fav]}$ as follows:

$$X_i^{[fav]} = X_i^{[fav]} \pm \left(X_i^{[fav]} \times fact_i\right) \tag{5.5}$$

where $fact_i = \begin{cases} random\,value \in \left(\sigma_1^{lower}, \sigma_1^{upper}\right) & if\ \frac{1}{X_i^{[fav]}} \leq \gamma \\ random\,value \in \left(\sigma_2^{lower}, \sigma_2^{upper}\right) & if\ \frac{1}{X_i^{[fav]}} > \gamma \end{cases}$ and $\sigma_1^{lower}, \sigma_1^{upper}$, $\sigma_2^{lower}, \sigma_2^{upper}$ are randomly generated values between 0 and 1 i.e., $0 < \sigma_1^{lower} < \sigma_1^{upper} \leq \sigma_2^{lower} < \sigma_2^{upper} < 1$. The value of γ as well as '+' or '−' sign in Eq. (5.5) are chosen based on the preliminary trials of the algorithm.

It gives a chance to every agent i to jump out of the local minima and may further help to search for a better solution. The perturbed solution is accepted if and only if the feasibility is maintained. Furthermore, every agent expands its sampling interval as follows:

$$\begin{gathered} \Psi_i \in \left[\left(\Psi_i^{lower} - \lambda_{up}\left\|\Psi_i^{upper} - \Psi_i^{lower}\right\|\right), \left(\Psi_i^{upper} + \lambda_{up}\left\|\Psi_i^{upper} - \Psi_i^{lower}\right\|\right)\right], \\ 0 < \lambda_{up} \leq 1 \end{gathered} \tag{5.6}$$

where λ_{up} is referred to as the interval factor corresponding to the expansion of sample space.

5.1.3 Modifications to Step 6 of the Unconstrained PC Approach

The detailed formulation of the modified convergence criterion is explained below and further presented in window C of the constrained PC algorithm flowchart in Fig. 5.1.

The current stable system objective $G(\mathbf{Y}^{[fav],s})$ and corresponding $\mathbf{Y}^{[fav],s}$ are accepted as the final solution referred to as $G(\mathbf{Y}^{[fav],final})$ and $\mathbf{Y}^{[fav],final} = \left\{X_1^{[fav],final}, X_2^{[fav],final}, \ldots, X_{N-1}^{[fav],final}, X_N^{[fav],final}\right\}$, if and only if either of the following conditions is satisfied.

If temperature $T = T_{final}$ or $T \to 0$.

If there is no significant change in the successive stable system objectives (i.e. $\left\|G(\mathbf{Y}^{[fav],s}) - G(\mathbf{Y}^{[fav],s-1})\right\| \leq \varepsilon$) for two successive implementations of the perturbation approach.

5.2 The Circle Packing Problem (CPP)

A generalized packing problem consists of determining how best to pack z objects into a predefined bounded space that yields best utilization of space with no overlap of object boundaries [13, 14]. The bounded space can also be referred to as a container. The packing objects and container can be circular, rectangular or irregular. Although the problem appears rather simple and in spite of its practical applications in production and packing for the textile, apparel, naval, automobile, aerospace, food industries, etc. [15] the CPP received considerable attention in the 'pure' mathematics literature but only limited attention in the operations research literature [16]. As it is proven to be a NP-hard problem [14, 17–19] and cannot be effectively solved by purely analytical approaches [20–30], a number of heuristic techniques were proposed solving the CPP [13, 14, 31–43]. Most of these approaches address the CPP in limited ways, such as close packing of fixed and uniform sized circles inside a square or circle container [14, 20–29], close packing of fixed and different sized circles inside a square or circle container [13, 15, 36–43], simultaneous increase in the size of the circles covering the maximum possible area inside a square [32–35], etc.

As per knowledge of the author of this book; the CPP was never solved in a distributed way. In this book, as a distributed MAS, every individual circle changes its size and position autonomously. This allows for addressing the important issue of the avoidance of the tragedy of commons which was also never addressed before in the context of the CPP. The next few sections describe the mathematical formulation and the solution to two cases of the CPP.

5.2.1 Formulation of the CPP

The objective of the CPP solved here was to cover the maximum possible area within a square by z number of circles without overlapping one another or exceeding the boundaries of the square. In order to achieve this objective, all the circles were allowed to increase their sizes as well as change their locations. The problem is formulated as follows:

$$\text{Minimize} \quad f = L^2 - \sum_{i=1}^{z} \pi r_i^2 \tag{5.7}$$

Subject to

$$\sqrt{(x_i - x_j)^2 + (y_i - y_j)^2} \geq r_i + r_j \tag{5.8}$$

$$x_i - r_i \geq x_l \tag{5.9}$$

$$x_i + r_i \leq x_u \tag{5.10}$$

$$y_i - r_i \geq y_l \tag{5.11}$$

$$y_i + r_i \leq y_u \tag{5.12}$$

$$0.001 \leq r_i \leq L/2 \tag{5.13}$$

$$i,j = 1, 2, \ldots, z \quad i \neq j \tag{5.14}$$

where

$L =$ length of the side of the square
$r_i =$ radius of circle i
$x_i,\ y_i =$ x and y coordinates of the center of circle i
$x_l,\ y_l =$ x and y coordinates of the lower left corner of the square
$x_u,\ y_u =$ x and y coordinates of the upper right corner of the square

In solving the proposed CPP using constrained PC approach presented in Sect. 5.1, the circles were considered as autonomous agents. These circles were assigned the strategy sets of coordinates and y coordinates of the center and the radius. Two cases of the CPP were solved. These cases differ from each other based on the initial configuration (location) of the circles as well as the way constraints are handled solving each case.

In Case 1, the circles were randomly initialized inside the square and were not allowed to cross the square boundaries. The constraints in Eq. (5.8) were satisfied using the Feasibility-based Rule I described in Sect. 5.1. And the constraints in Eqs. (5.9–5.12) were satisfied in every iteration of the algorithm using a repair approach. The repair approach refers to pushing the circles inside the square if they crossed the boundaries of it. It is similar to the one proposed in Chap. 3 solving the MTSP. In Case 2, the circles were randomly located in-and-around the square and all the constraints from (5.8) to (5.12) were satisfied using the Feasibility-based Rule I described in Sect. 5.1. The initial configuration of Case 1 and Case 2 is shown in Figs. 5.2a and 5.5a, respectively.

The constrained PC algorithm solving both the cases was coded in MATLAB 7.8.0 (R2009A) and the simulations were run on a Windows platform using an Intel Core 2 Duo, 3 GHz processor speed and 3.25 GB memory capacity. Furthermore, for both the cases the set of parameters chosen was as follows: (a) individual agent

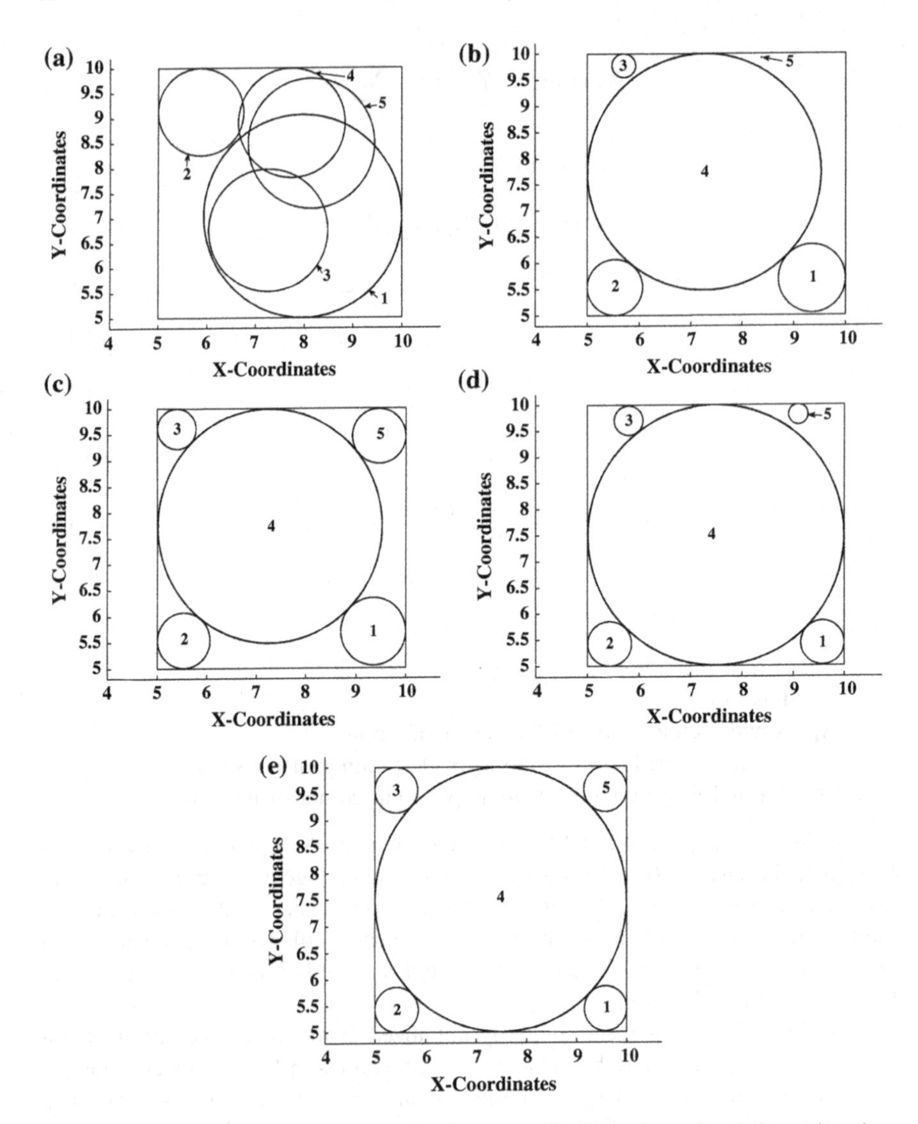

Fig. 5.2 Solution history for case 1 **a** randomly generated initial solution **b** solution at iteration 401 **c** solution at iteration 901 **d** solution at iteration 1,001 **e** stable solution at iteration 1055

sample size $m_i = 5$, (b) number of test iterations $n_{test} = 20$, (c) the shrinking interval factor $\lambda_{down} = 0.05$, (d) the expansion interval factor $\lambda_{up} = 0.1$, (e) perturbation parameters $\sigma_1^{lower} = 0.001$, $\sigma_1^{upper} = 0.01$, $\sigma_2^{lower} = 0.5$, $\sigma_2^{upper} = 0.7$, $\gamma = 0.99$ and the sign in Eq. (5.5) was chosen to be '−'. In addition to it, a voting heuristic was also incorporated in the constrained PC algorithm. It is described in the Sect. 5.2.4.

5.2.2 *Case 1: CPP with Circles Randomly Initialized Inside the Square*

In this case of the CPP, five circles ($z = 5$) were initialized randomly inside the square without exceeding the boundary edges of the square. The length of the side of the square was five units (i.e. $L = 5$). More than 30 runs of the constrained PC algorithm described in Sect. 5.1 were conducted solving the Case 1 of the CPP with different initial configurations of the circles inside the square. The true optimum solution was achieved in every run with the average CPU time of 14.05 min and average number of function evaluations is 17,515.

The randomly generated initial solution, the intermediate iteration solutions, and the converged true optimum solution from one of the instances are presented in Fig. 5.2. The corresponding convergence plot of the system objective is presented in Fig. 5.3. The convergence of the associated variables such as radius of the circles, x coordinates and y coordinates of the center of the circles is presented in Fig. 5.4a–c, respectively. The solution was converged at iteration 1035 with 26910 function evaluations. The true optimum value of the objective function (f) achieved was 3.0807 units.

As mentioned before, the algorithm was assumed to have converged when successive implementations of the perturbation approach stabilize to equal objective function value. It is evident from Figs. 5.2, 5.3 and 5.4 that the solution was converged to true optimum at iteration 1035 as the successive implementations of the perturbation approach produced stable and equal objective function values. Furthermore, it is also evident from Figs. 5.3 and 5.4 that the solution was perturbed at iteration 778, 901, 1136 and 1300. It is clear that the implementation of the perturbation approach at iteration 901 helped the solution to jump out of the local minima and further achieve the true optimum solution at iteration 1035.

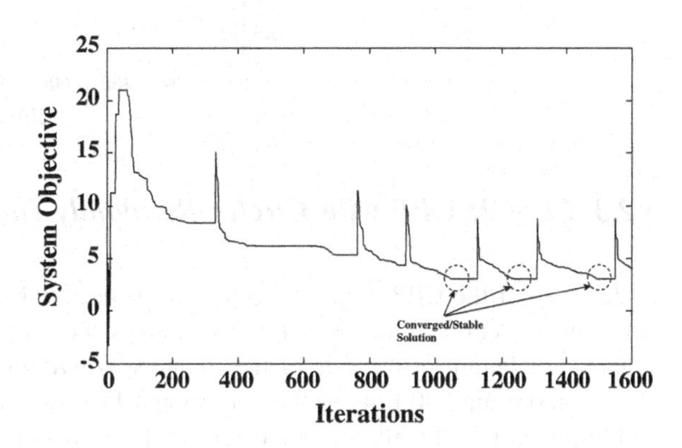

Fig. 5.3 Convergence of the objective function for case 1

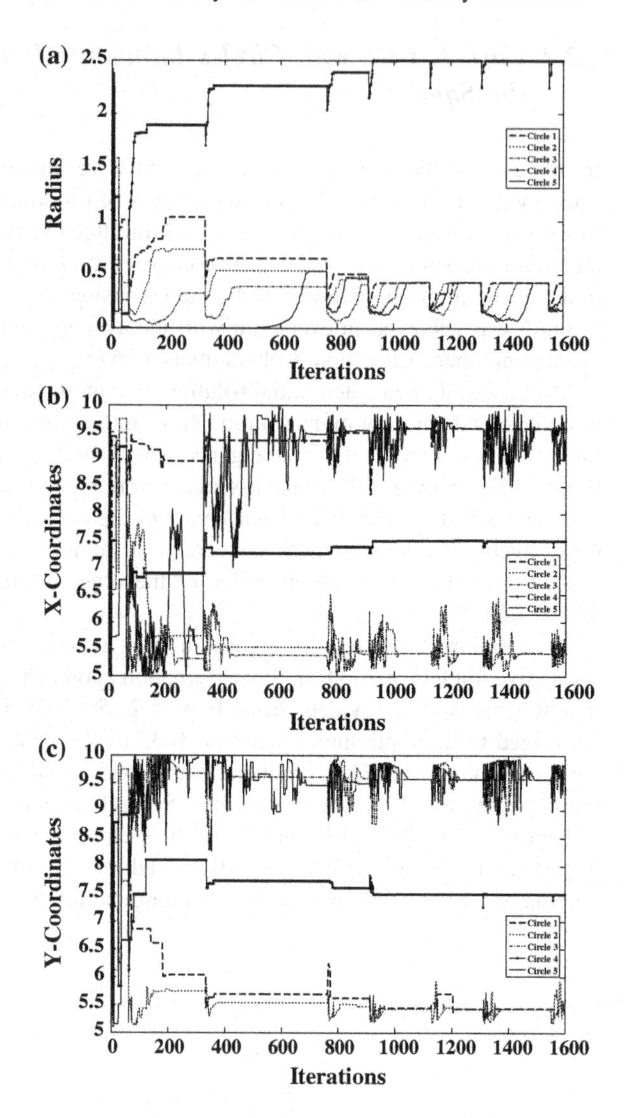

Fig. 5.4 Convergence of the strategies for case 1 **a** convergence of the radius **b** convergence of the x-coordinates of the center **c** convergence of the y-coordinates of the center

5.2.3 Case 2: CPP with Circles Randomly Initialized

In this case of the CPP, five circles ($z = 5$) were initialized randomly in the space with no restriction as in Case 1 where circles were randomly placed inside the square. The length of the side of the square was five units (i.e. $L = 5$). Similar to Case 1, more than 30 runs of the constrained PC algorithm described in Sect. 5.1 with different initial configuration of the circles were conducted solving the Case 2. The true optimum solution was achieved in every run with the average CPU time of 14.05 min and average number of function evaluations is 68,406.

The randomly generated initial solution, the intermediate iteration solutions, and the converged true optimum solution from one of the instances of Case 2 are presented in Fig. 5.5. The corresponding convergence plot of the system objective is presented in Fig. 5.6. The convergence of the associated variables such as radius of the circles, x coordinates and y coordinates of the center of the circles is presented in Fig. 5.7a–c, respectively. The solution was converged at iteration 955 with 24,830 function evaluations. The true optimum value of the objective function (f) achieved was 3.0807 units.

In the instance of the Case 2 presented here, the solution was perturbed at iteration 788, 988, 1170 and 1355. It is clear that the implementation of the perturbation approach at iteration 788 helped the solution to jump out of the local minima and further achieve the true optimum solution at iteration 955. It is important to mention that the instance illustrated here did not require the voting heuristic to be applied.

5.2.4 Voting Heuristic

In a few instances of the CPP cases solved here, in order to jump out of the local minimum, a voting heuristic was required. It was implemented in conjunction with the perturbation approach. Once the solution was perturbed, every circle voted 1 for each quadrant which it does not belong to at all, and voted 0 otherwise. The circle with the smallest size shifted itself to the extreme corner of the quadrant with the highest number of votes, i.e. the winner quadrant. The new position of the smallest size circle was confirmed only when the solution remained feasible and the algorithm continues. If all the quadrants acquire equal number of votes, no circle moves its position and the algorithm continues. The voting heuristic is demonstrated in Fig. 5.8.

A voting grid corresponding to every quadrant of the square in Fig. 5.8a is represented in Fig. 5.8b. The solid circles represent the solution before perturbation while corresponding perturbed ones are represented in dotted lines. The votes given by the perturbed circles (dotted circles) to the quadrants are presented in the grid. As the maximum number of votes are given to quadrant 1 (i.e. Q 1), the circle with smallest size (circle 3) shifts to the extreme corner of the quadrant Q 1 and confirms the new position as the solution remains feasible. Based on the trials conducted so far, it was noticed that the voting heuristic was not necessary to be implemented in every run of the constrained PC algorithm solving the CPP. Moreover, in those of the few runs in which the voting heuristic was required, it was required to be implemented only once in the entire execution of the algorithm. A variant of the voting heuristic was also implemented in conjunction with energy landscape paving algorithm [15, 37, 38] in which, the smallest circle was picked and placed randomly

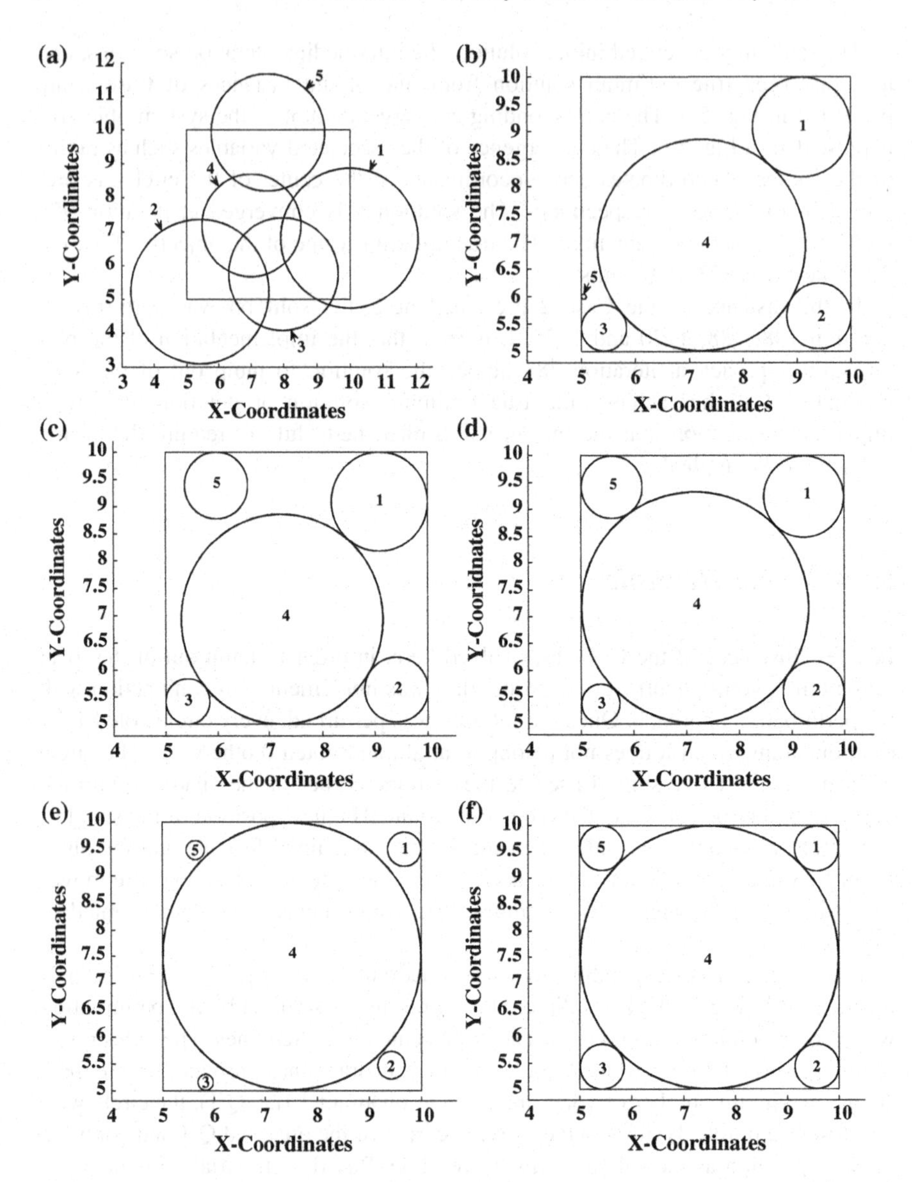

Fig. 5.5 Solution history for case 2 **a** randomly generated initial solution **b** solution at iteration 301 **c** solution at iteration 401 **d** solution at iteration 601 **e** solution at iteration 801 **f** stable solution at iteration 955

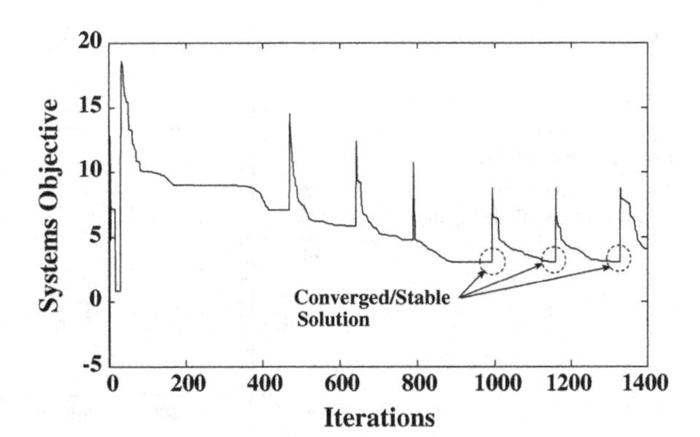

Fig. 5.6 Solution convergence plot for case 2

at the vacant place to produce a new configuration. It was claimed that this heuristic helped the algorithm jump out of the local minima. Furthermore, this heuristic was required to be implemented in every iteration of the algorithm.

5.2.5 Agent Failure Case

As discussed before, a centralized system is vulnerable to single point failures and its performance may be severely affected if an agent in the system fails. On the other hand, in case of a distributed and decentralized algorithm, the system is more robust as the system is controlled by its autonomous subsystems.

As mentioned in Sect. 2.3, the failed agent in PC can be considered as the one that does not communicate with other agents in the system and does not update its probability distribution. This does not prevent other agents from continuing further, by simply considering the failed agent's latest communicated strategies as the current strategies.

The immunity to agent failure is essential in the field of UAV formation and collaborative path planning because failure of the centralized controller would be devastating and may result in collision. This is addressed in the related work on UAVs [44–51].

In the context of the CPP, as a variation of Case 1, circle 2 was failed at a randomly chosen iteration 30, i.e. circle 2 does not update its x and y coordinates as well as its radius after iteration 30. More than 30 cases of the constrained PC algorithm with Feasibility-based Rule I were conducted solving the CPP with agent failure case and the average function evaluations were 17,365. For the case presented here, the number of function evaluations was 19,066 and the randomly generated initial solution, the intermediate iteration solutions, and the converged true optimum solution are shown in Fig. 5.9. The corresponding convergence plot

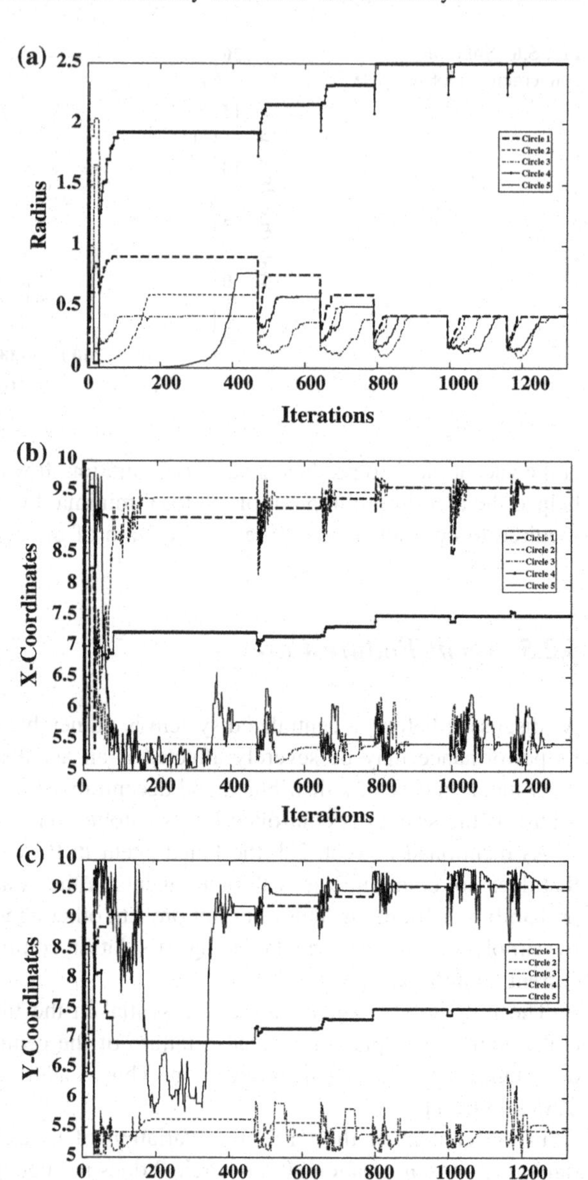

Fig. 5.7 Convergence of the strategies for case 2
a convergence of the radius
b convergence of the x-coordinates of the center
c convergence of the y-coordinates of the center

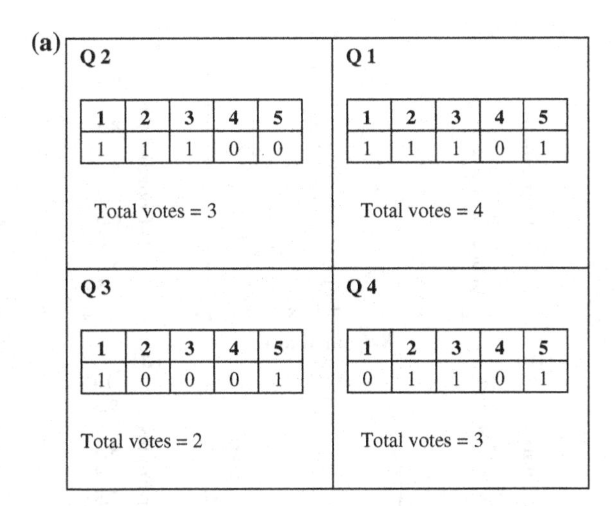

(a)
Q 2
1 2 3 4 5
1 1 1 0 0
Total votes = 3
Q 1
1 2 3 4 5
1 1 1 0 1
Total votes = 4
Q 3
1 2 3 4 5
1 0 0 0 1
Total votes = 2
Q 4
1 2 3 4 5
0 1 1 0 1
Total votes = 3

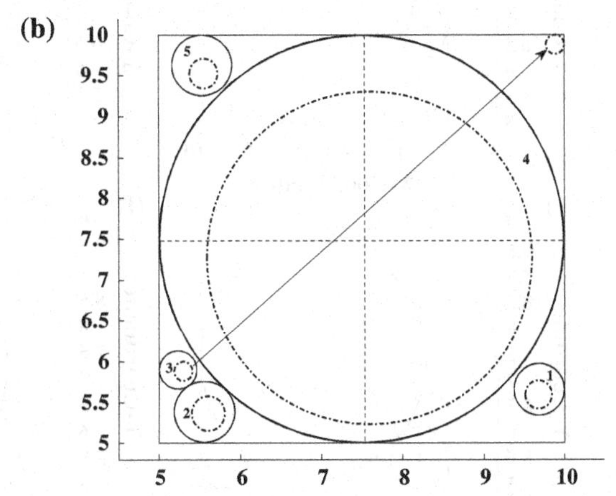

Fig. 5.8 Voting heuristic **a** a case for voting heuristic **b** voting grids

of the system objective is presented in Fig. 5.10. The convergence of the associated variables such as radius of the circles, x coordinates and y coordinates of the center of the circles is presented in Fig. 5.11a–c, respectively.

It is worth to mention that the voting heuristic was not required in the case presented in Fig. 5.9. This is because according to the voting heuristic the smallest circle which moves itself to the new position is itself the failed agent. In addition, the same set of parameter as values listed in Sect. 5.2.1 was used for the agent failure case.

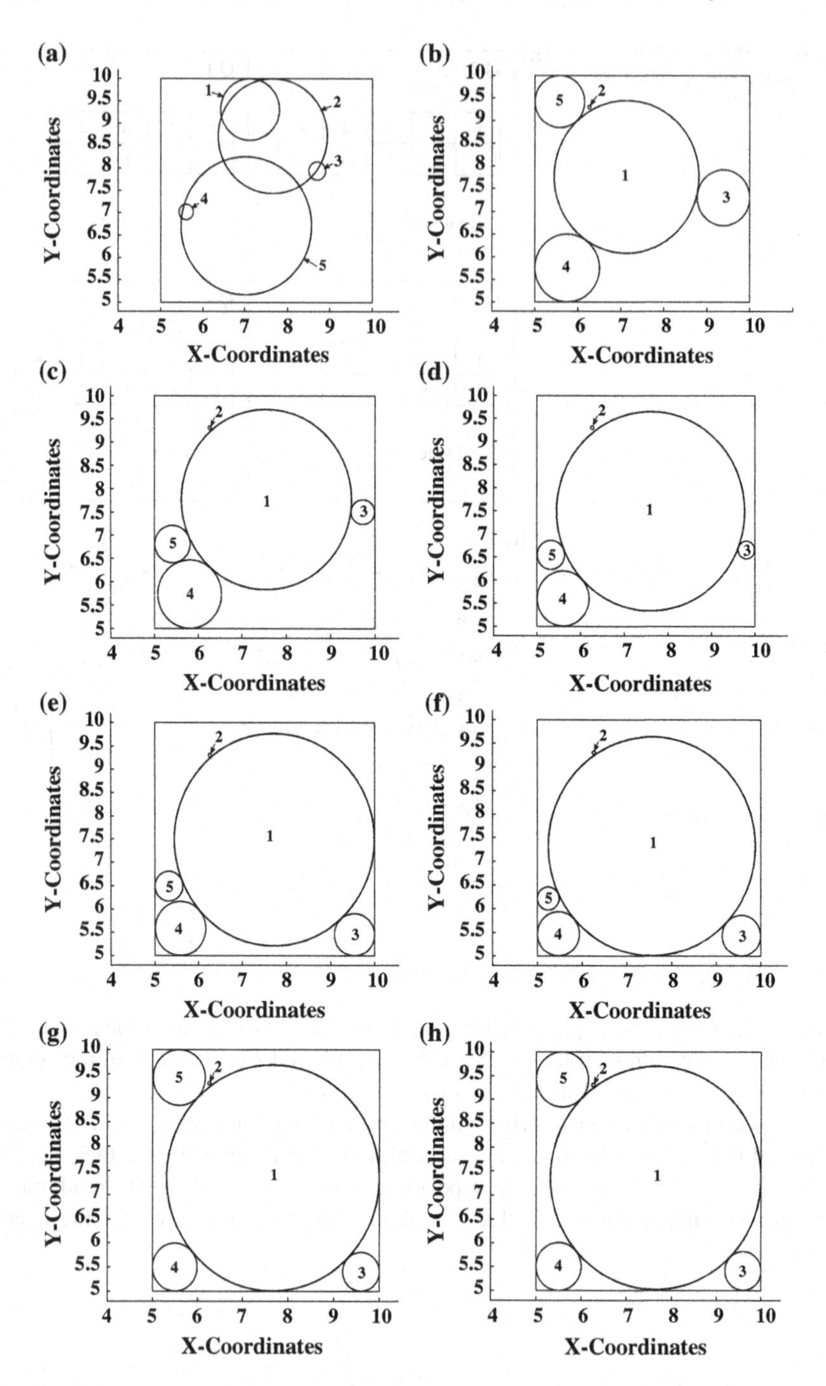

Fig. 5.9 Solution history for agent failure case **a** randomly generated iteration solution **b** solution at iteration 124 **c** solution at iteration 231 **d** solution at iteration 377 **e** solution at iteration 561 **f** solution at iteration 723 **g** stable solution at iteration 901 **h** stable solution at iteration 1051

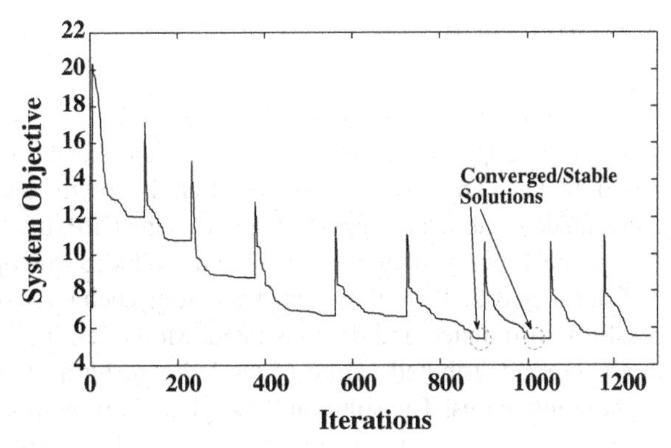

Fig. 5.10 Convergence of the objective function for agent failure case

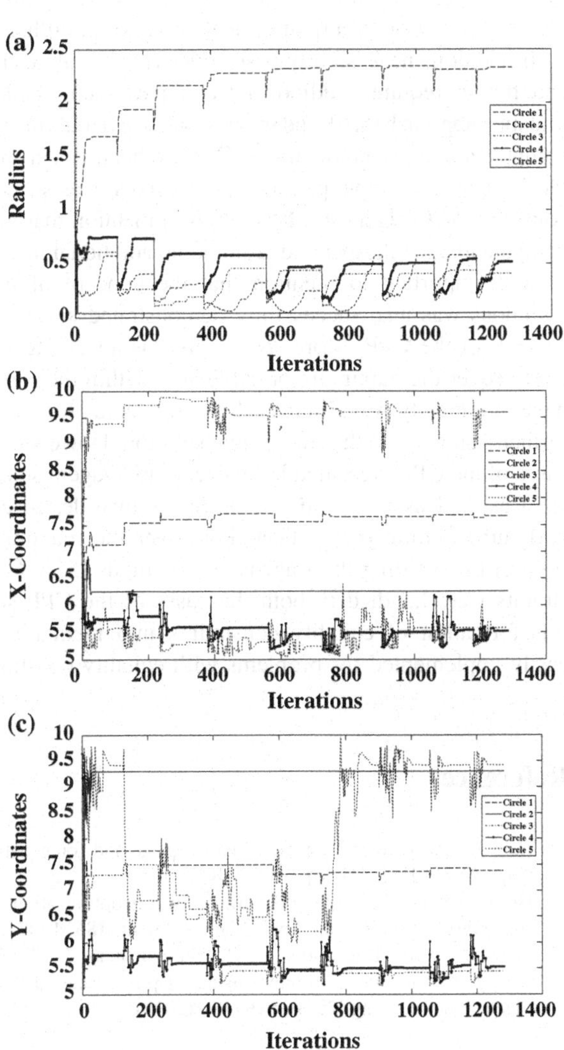

Fig. 5.11 Convergence of the strategies for agent failure case **a** convergence of the radius **b** convergence of the x-coordinates of the center **c** convergence of the y-coordinates of the center

5.3 Discussion

The above sections described the successful implementation of a generalized constrained PC approach using a variation of the feasibility-based rule originally proposed in [1]. The results indicated that it could successfully be used to solve constrained optimization problems such as the CPP. It is evident from the results that the approach was sufficiently robust and produced true optimum results in every run of Case 1 and 2. It implies that the rational behavior of the agents could be successfully formulated and demonstrated. Moreover, it is also evident that because of the inherent distributed nature of the PC algorithm, it can easily accommodate the agent failure cases. The solution highlights its strong potential to deal with the agent failure which may arise in real world complex problems including urban traffic control, formation of airplanes fleet and mid-air collision avoidance, etc.

In addition, the feasibility-based rule in [1–7] suffered from maintaining the diversity and further required additional techniques such as niching [1], SA [2], modified mutation approach [5, 6], and several associated trials in [3–6], etc. It may require further computations and memory usage. On the other hand, in order to jump out of the possible local minima, a simple perturbation approach was successfully incorporated into the constrained PC algorithm. It is worth to mention that the perturbation approach was computationally cheaper and required no additional memory usage.

It is important to mention that the concept of the avoidance of tragedy of commons was also successfully demonstrated solving the two cases (Case 1 and Case 2) of the CPP. More specifically, as all the agents were supposed to collectively cover the maximum possible area within the square, every circle/agent could have in a greedy way, increased its own radius (or size) [11, 12]. This would have further resulted into the suboptimal solution. However, in the PC solution to the two cases of the CPP presented here, every circle/agent selected its individual radius (or size) as well as its x and y coordinates in order to collectively achieve the true optimum solution (i.e. collectively cover the maximum possible area within the square) by avoiding the tragedy of commons [52]. Although only inequality constraints were handled in both the cases of the CPP solved here, the approach of transformation of equality constraints into inequality constraints [8–10] can be easily implemented for problems with equality constraints.

References

1. Deb, K.: An efficient constraint handling method for genetic algorithms. Comput. Methods Appl. Mech. Eng. 186, pp. 311–338 (2000)
2. He, Q., Wang, L.: A hybrid particle swarm optimization with a feasibility-based rule for constrained optimization. Appl. Math. Comput. **186**, 1407–1422 (2007)
3. Sakthivel, V.P., Bhuvaneswari, R., Subramanian, S.: Design optimization of three-phase energy efficient induction motor using adaptive bacterial foraging algorithm. Comput. Math. Electr. Electron. Eng. **29**(3), 699–726 (2010)

4. Kaveh, A., Talatahari, S.: An improved ant colony optimization for constrained engineering design problems. Comput. Aided Eng. Software **27**(1), 155–182 (2010)
5. Bansal, S., Mani, A., Patvardhan, C.: Is stochastic ranking really better than feasibility rules for constraint handling in evolutionary algorithms? In: Proceedings of World Congress on Nature and Biologically Inspired Computing, pp. 1564–1567 (2009)
6. Gao, J., Li, H., Jiao, Y.C.: Modified differential evolution for the integer programming problems. In: Proceedings of International Conference on Artificial Intelligence, pp. 213–219 (2009)
7. Ping, W., Xuemin, T.: A Hybrid DE-SQP algorithm with switching procedure for dynamic optimization. In: Proceedings of Joint 48th IEEE Conference on Decision and Control and 28th Chinese Control Conference, pp. 2254–2259 (2009)
8. Ray, T., Tai, K., Seow, K.C.: An evolutionary algorithm for constrained optimization. In: Proceedings of the Genetic and Evolutionary Computation Conference, pp. 771–777 (2000)
9. Ray, T., Tai, K., Seow, K.C.: Multiobjective design optimization by an evolutionary algorithm. Eng. Optim. **33**(4), 399–424 (2001)
10. Tai, K., Prasad, J.: Target-matching test problem for multiobjective topology optimization using genetic algorithms. Struct. Multi. Optim. **34**(4), 333–345 (2007)
11. Kulkarni, A.J., Tai, K.: A probability collectives approach with a feasibility-based rule for constrained optimization. Appl. Comput. Intell. Soft Comput. 2011, Article ID 980216
12. Metkar, S.J., Kulkarni, A.J.: Boundary searching genetic algorithm: a multi-objective approach for constrained problems. In: Satapathy, S.C., Biswal, B.N., Udgata, S.K. (eds) Advances in Intelligent and Soft Computing, Springer, pp. 269–276 (2014)
13. Zhang, D., Deng, A.: An effective hybrid algorithm for the problem of packing circles into a larger containing circle. Comput. Oper. Res. **32**, 1941–1951 (2005)
14. Theodoracatos, V.E., Grimsley, J.L.: The optimal packing of arbitrarily-shaped polygons using simulated annealing and polynomial-time cooling schedules. Comput. Methods Appl. Mech. Eng. **125**, 53–70 (1995)
15. Liu, J., Xue, S., Liu, Z., Xu, D.: An improved energy landscape paving algorithm for the problem of packing circles into a larger containing circle. Comput. Ind. Eng. **57**(3), 1144–1149 (2009)
16. Castillo, I., Kampas, F.J., Pinter, J.D.: Solving circle packing problems by global optimization: numerical results and industrial applications. Eur. J. Oper. Res. **191**(3), 786–802 (2008)
17. Garey, M.R., Johnson D.S.: Computers and Intractability: A Guide to the Theory of NP-completeness. W. H. Freeman & Co. (1979)
18. Hochbaum, D.S., Maass, W.: Approximation schemes for covering and packing problems in image processing and VLSI. J. Assoc. Comput. Mach. **1**(32), 130–136 (1985)
19. Wang, H., Huang, W., Zhang, Q., Xu, D.: An improved algorithm for the packing of unequal circles within a larger containing circle. Eur. J. Oper. Res. **141**, 440–453 (2002)
20. Szabo, P.G., Markot, M.C., Csendes, T.: Global optimization in geometry—circle packing into the square. Essays and Surveys. In: Audet, P., Hansen, P., Savard, G. (eds.) Global Optimization, Kluwer , pp. 233–265 (2005)
21. Nurmela, K.J., Ostergard, P.R.J.: Packing up to 50 equal circles in a square. Discrete Comput. Geom. **18**, 111–120 (1997)
22. Nurmela, K.J., Ostergard, P.R.J.: More optimal packing of equal circles in a square. Discrete Comput. Geom. **22**, 439–457 (1999)
23. Graham, R.L., Lubachevsky, B.D.: Repeated patterns of dense packings of equal disks in a square. Electr. J. Combinatorics **3**, 1–16 (1996)
24. Szabo, P.G., Csendes, T., Casado, L.G., Garcia, I.: Equal circles packing in a square i—problem setting and bounds for optimal solutions. In: Giannessi, F., Pardalos, P., Rapcsak,

T. (eds.) Optimization Theory: Recent Developments from Matrahaza, Kluwer, pp. 191–206 (2001)
25. de Groot, C., Peikert, R., Wurtz, D.: The Optimal Packing of Ten Equal Circles in a Square, IPS Research Report 90-12. ETH, Zurich (1990)
26. Goldberg, M.: The packing of equal circles in a square, Mathematics Magazine, 43 pp. 24–30 (1970)
27. Mollard, M., Payan, C.: Some progress in the packing of equal circles in a square. Discrete Math. **84**, 303–307 (1990)
28. Schaer, J.: On the packing of ten equal circles in a square. Math. Mag. **44**, 139–140 (1971)
29. Schluter, K.: Kreispackung in quadraten. Elem. Math. **34**, 12–14 (1979)
30. Valette, G.: A better packing of ten circles in a square. Discrete Math. **76**, 57–59 (1989)
31. Boll, D.W., Donovan, J., Graham, R.L., Lubachevsky, B.D.: Improving dense packings of equal disks in a square. Electr. J. Combinatorics **7**, R46 (2000)
32. Lubachevsky, D., Graham, R.L.: Curved hexagonal packing of equal circles in a circle. Discrete Comput. Geom. **18**, 179–194 (1997)
33. Szabo, P. G. Specht, E.: Packing up to 200 equal circles in a square. In: Torn, A., Zilinskas, J. (eds.) Models And Algorithms For Global Optimization, pp. 141–156 (2007)
34. Mladenovic, N., Plastria, F., Urosevi, D.: Formulation space search for circle packing problems. LNCS, pp. 212–216 (2007)
35. Mldenovic, N., Plastria, F., Urosevi, D.: Reformulation descent applied to circle packing problems. Comput. Oper. Res. **32**, 2419–2434 (2005)
36. Huang, W., Chen, M.: Note on: an improved algorithm for the packing of unequal circles within a larger containing circle. Comput. Ind. Eng. **50**(2), 338–344 (2006)
37. Liu, J., Xu, D., Yao, Y., Zheng, Y.: Energy landscape paving algorithm for solving circles packing problem. In: International Conference on Computational Intelligence and Natural Computing, pp. 107–110 (2009)
38. Liu, J., Yao,Y., Zheng, Y., Geng, H., Zhou, G.: An effective hybrid algorithm for the circles and spheres packing problems. LNCS-5573, pp. 135–144 (2009)
39. Stoyan, Y.G., Yaskov, G.N.: Mathematical model and solution method of optimization problem of placement of rectangles and circles taking into account special constraints. Int. Trans. Oper. Res. **5**, 45–57 (1998)
40. Stoyan, Y.G., Yaskov, G.N.: A mathematical model and a solution method for the problem of placing various-sized circles into a strip. Eur. J. Oper. Res. **156**, 590–600 (2004)
41. George, J.A.: Multiple container packing: a case study of pipe packing. J. Oper. Res. Soc. **47**, 1098–1109 (1996)
42. George, J.A., George, J.M., Lamar, B.W.: Packing different-sized circles into a rectangular container. Eur. J. Oper. Res. **84**, 693–712 (1995)
43. Hifi, M., M'Hallah, R.: Approximate algorithms for constrained circular cutting problems. Comput. Oper. Res. **31**, 675–694 (2004)
44. Bieniawski, S.R.: Distributed optimization and flight control using collectives. Ph.D. Dissertation, Stanford University, CA, USA (2005)
45. Chandler, P., Rasumussen, S., Pachter, M.: UAV cooperative path-planning. In: Proceedings of AIAA Guidance, Navigation, and Control Conference. Paper. No. 4370 (2000)
46. Sislak, D., Volf, P., Komenda, A., Samek, J., Pechoucek, M.: Agent-based multi-layer collision avoidance to unmanned aerial vehicles. In: Proceedings of International Conference on Integration of Knowledge Intensive Multi-Agent Systems, Art. No. 4227576, pp. 365–370 (2007)
47. Krozel, J., Peters, M., Bilimoria, K.: A decentralized control strategy for distributed air/ground traffic separation. In: Proceedings of AIAA Guidance, Navigation and Control Conference and Exhibit, Paper, No. 4062 (2000)

48. Zengin, U., Dogan, A.: Probabilistic trajectory planning for UAVs in dynamic environments. In: Proceedings of AIAA 3rd "Unmanned-Unlimited" Technical Conference, Workshop, and Exhibit 2, Paper, No. 6528 (2004)
49. Anderson, M.R., Robbins, A.C.: Formation flight as a cooperative game. In: Proceedings of AIAA Guidance, Navigation, and Control Conference, pp. 244–251 (1998)
50. Chang, D.E., Shadden, Marsden J. E., Olfati-Saber, R.: Collision avoidance for multiple agent systems. In: Proceedings of 42nd IEEE Conference on Decision and Control, pp. 539–543 (2003)
51. Sigurd, K., How, J.: UAV trajectory design using total field collision avoidance. In: Proceedings of AIAA Guidance, Navigation, and Control Conference, Paper No. 5728 (2003)
52. Haddin, G.: The tragedy of commons. Science **162**, 1243–1248 (1968)

Chapter 6
Probability Collectives for Discrete and Mixed Variable Problems

This chapter demonstrates the ability of PC for solving practically important discrete and mixed variable problems in structural and mechanical engineering domain. The truss structure problems such as 17-bar, 25-bar, 72-bar, 45-Bar, 10-Bar and 38-Bar, a helical compression spring design, a reinforced concrete beam design, stepped cantilever beam design and speed reducer were successfully solved. The Feasibility-based Rule I discussed in Chap. 5 was applied for handling the constraints. Furthermore, the perturbation approach assisting the Feasibility-based Rule I was not required for solving these problems which also avoided several associated parameters to be tuned. The results were comparable with the existing state-of-the-art methods.

These problems are well studied in the literature and used to compare the performance of various optimization algorithms [1–30] such as Genetic Algorithm (GA) [1, 2], Harmony Search (HS) [3, 4], Particle Swarm Optimization (PSO), PSO with Passive Congregation (PSOPC), Hybrid PSO [5, 6], Discrete Heuristic Particle Swarm Ant Colony Optimization (DHPSACO) [7], Genetic Adaptive Search (GeneAS) [8], Steady state GA (SGA) [1], Penalty-based GA [9], Genetic Algorithm Based Optimum Structural Design Approach (GAOS) [13], Meta GA (MGA) [10], Adaptive Hybrid GA (AHGA) [11], Parameter-less Adaptive Penalty Scheme with GA (GA-APM) [12], hybridization of GA with an Artificial Immune System (AIS-GA) and AIS-GA with clearing approach (AIS-GA-C) [14], a GA-AIS hybridized approach [15], Mutation-based Real-coded Genetic Algorithm (MBRCGA) [16], Branch and Bound (B&B) Method [17], Hybrid Swarm Intelligence Approach (HSIA) [18], Mine Blast Algorithm (MBA) [19], Firefly Algorithm (FA) [20], Generalized Hopfield Network based Augmented Lagrange Multiplier approach GHN-ALM and GHN based Extended Penalty Approach GHN-EP [21], Nonlinear B&B Method [22], Discrete Steepest Descent and Rotating Coordinate Direction Methods (SD-RC) [23], Evolutionary Algorithms (EA) [24] Differential EA (DEA) [25] Rank-Niche Evolution Strategy (RNES) [26], Bacterial Foraging Behavior Algorithm (BFBA) [27], Nonlinear B&B Method with Round up (B&B-RU) Precise Discrete (B&B-PD), Linear Approximate Discrete (B&B-LAD) and Conservative Approximate Discrete (B&B-CAD) [28]. In addition, few heuristics such as multi-start continuous local optimization and rounding

© Springer International Publishing Switzerland 2015

A.J. Kulkarni et al., *Probability Collectives*, Intelligent Systems Reference Library 86, DOI 10.1007/978-3-319-16000-9_6

[29], and gradient based approach integrated with a combinatorial loop [30] were also specifically developed for solving structural optimization problems.

The constrained PC algorithm was coded in MATLAB 7.7.0 (R2008b) and the simulations were run on Windows platform using an Intel Core i5, 2.8 GHz processor speed and 4 GB RAM [31–33]. The mathematical formulation, results and comparison of solved problems with other contemporary algorithms are discussed below.

Problem 6.1: 17-Bar Truss Structure

This problem was solved in [2, 3, 5] having 17-bar shown in Fig. 6.1. The aim is to minimize the weight f subject to stress and deflection constraints.

$$\text{Minimize } f = W = \sum_{i=1}^{N} \rho A_i l_i \tag{6.1}$$

$$\text{Subject to } |\sigma_i| \le \sigma_{\max} \quad i = 1, 2, 3, \ldots, N \tag{6.2}$$

$$|u_j| \le u_{\max} \quad j = 1, 2, 3, \ldots, M$$

The weight density of material ρ is 0.268 lb/in^3 and modulus of elasticity E is 30,000 ksi. The members are subjected to stress limitations $\sigma_{\max}$ of ± 50 ksi and displacement limitations $u_{\max}$ of ± 2.0 in are imposed on all nodes in maximum allowable stress in both directions (x and y). The single vertical downward load of 100 kips at node 9 was considered. There are seventeen independent design variables. The minimum cross-sectional area of the members is 0.1 in^2.

This problem was solved in [2, 3, 5] as a continuous variables problem. The results obtained from these approaches along with PC approach were listed in Table 6.1. The PC solution to the problem produced competent results with reasonable computational cost. The best, mean and worst function values found from twenty trials performed were 2584.0255, 2586.1132 and 2589.2534, respectively with standard deviation 1.375508. The average CPU time, average number of

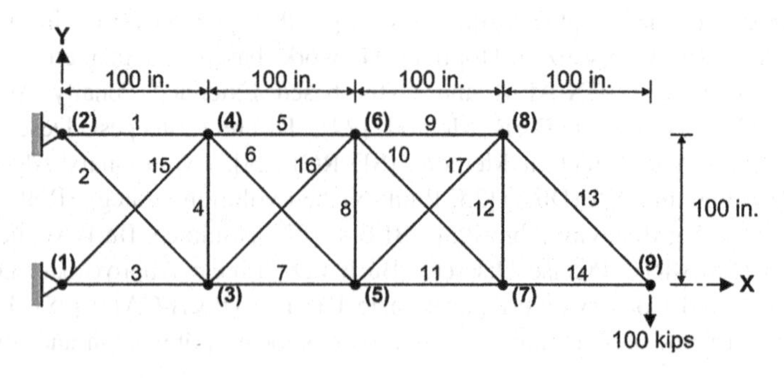

Fig. 6.1 17-Bar truss structure

Table 6.1 Performance comparison of various algorithms solving 17-bar truss structure problem

Design variables	GA [2]	HS [3]	PSO [5]	PSOPC [5]	HPSO [5]	PC
x_1	16.029	15.821	15.766	15.981	15.896	15.6498
x_2	0.107	0.107	2.263	0.100	0.103	0.1788
x_3	12.183	11.996	13.854	12.142	12.092	12.3751
x_4	0.110	0.100	0.106	0.100	0.100	0.109
x_5	8.417	8.150	11.356	8.098	8.063	8.3895
x_6	5.715	5.507	3.915	5.566	5.591	5.3713
x_7	11.331	11.829	8.071	11.732	11.915	12.1696
x_8	0.105	0.100	0.100	0.100	0.100	0.1138
x_9	7.301	7.934	5.850	7.982	7.965	7.8897
x_{10}	0.115	0.100	2.294	0.113	0.100	0.1074
x_{11}	4.046	4.093	6.313	4.074	4.076	3.9733
x_{12}	0.101	0.100	3.375	0.132	0.100	0.1247
x_{13}	5.611	5.660	5.434	5.667	5.670	5.4795
x_{14}	4.046	4.061	3.918	3.991	3.998	4.1713
x_{15}	5.152	5.656	3.534	5.555	5.548	5.5829
x_{16}	0.107	0.100	2.314	0.101	0.103	0.1548
x_{17}	5.286	5.582	3.542	5.555	5.537	5.3948
$f(lb)$	2594.42	2580.81	2724.37	2582.85	2582.85	2584.0255

function evaluations and associated parameters are listed in Table 6.21. The convergence plot for best PC solution is presented in Fig. 6.2.

Problem 6.2: 25-Bar Truss Structure

The 25 bar 3-D truss structure problem [1, 4, 6, 7] shown in Fig. 6.3 has aim to minimize the weight f (or W) subject to minimum stress and minimum deflection as follows:

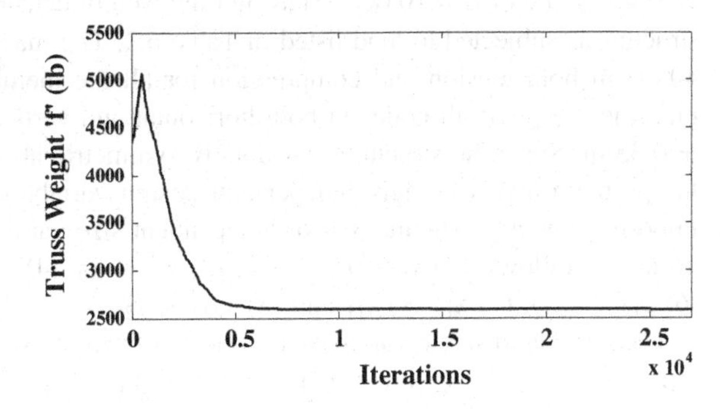

Fig. 6.2 Convergence plot for minimum weight of 17-bar truss structure problem

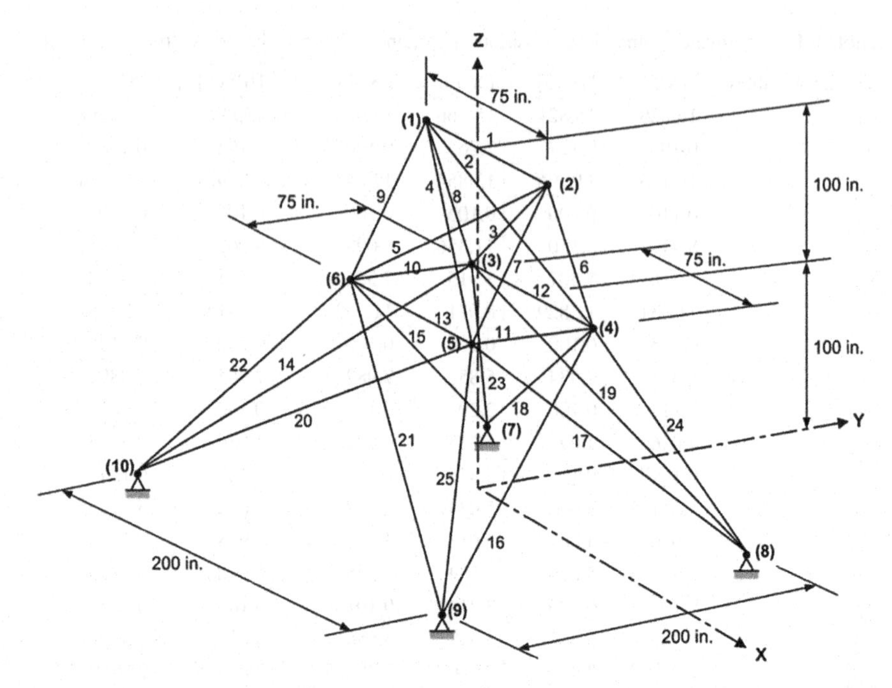

Fig. 6.3 25-Bar truss structure

$$\text{Minimize } f = W = \sum_{i=1}^{N} \rho A_i l_i \tag{6.3}$$

$$\text{Subject to } |\sigma_i| \leq \sigma_{\max} \quad i = 1, 2, 3, \ldots, N \tag{6.4}$$

$$|u_j| \leq u_{\max} \quad j = 1, 2, 3, \ldots, M$$

All truss members were assumed to be constructed from a material with an elastic module of E = 10,000 ksi x_5 and the weight density of $\rho = 0.1\ lb/\text{in}^3$. The structure is subjected to load listed in Table 6.2. The maximum stress limit $\sigma_{\max}$ is 40 ksi in both tension and compression for all the members. The maximum displacement $u_{\max}$ of all nodes in both horizontal and vertical directions is limited to $\pm$ 0.35 in. Since the structure was doubly symmetric about the x-axis and y-axes, the problem involved eight independent design variables after linking in order to impose symmetry. The number of independent size variables was reduced to eight groups as follows: (1) x_1, (2) $x_2 \sim x_5$, (3) $x_6 \sim x_9$, (4) $x_{10} \sim x_{11}$, (5) $x_{12} \sim x_{13}$, (6) $x_{14} \sim x_{17}$, (7) $x_{18} \sim x_{21}$ and (8) $x_{22} \sim x_{25}$.

Case 1: The discrete variables are selected from the set x_i = {0.01, 0.4, 0.8, 1.2, 1.6, 2.0, 2.4, 2.8, 3.2, 3.6, 4.0, 4.4, 4.8, 5.2, 5.6, 6.0} in^2.

Table 6.2 Loading condition for 25 bar

Nodes	P_x	P_y *kips*	P_z *kips*	P_x *kips*	P_y *kips*	P_z *kips*
1	0.0	20.0	−5.0	1.0	10.0	−5.0
2	0.0	−20.0	−5.0	0.0	10.0	−5.0
3	0.0	0.0	0.0	0.5	0.0	0.0
6	0.0	0.0	0.0	0.5	0.0	0.0

Table 6.3 Available cross section area of the AISC code

Sr no	in^2	Sr no	in^2	Sr no	in^2
1	0.111	23	2.62	45	7.97
2	0.141	24	2.63	46	8.53
3	0.196	25	2.88	47	9.3
4	0.25	26	2.93	48	10.85
5	0.307	27	3.09	49	11.5
6	0.391	28	3.13	50	13.5
7	0.443	29	3.38	51	13.9
8	0.563	30	3.47	52	14.2
9	0.602	31	3.55	53	15.5
10	0.766	32	3.63	54	16
11	0.785	33	3.84	55	16.9
12	0.994	34	3.87	56	18.8
13	1	35	3.88	57	19.9
14	1.228	36	4.18	58	22
15	1.266	37	4.22	59	22.9
16	1.457	38	4.49	60	24.5
17	1.563	39	4.59	61	26.5
18	1.62	40	4.8	62	28
19	1.8	41	4.97	63	30
20	1.99	42	5.12	64	33.5
21	2.13	43	5.74	–	–
22	2.38	44	7.22	–	–

Case 2: The discrete variables are selected from the American Institute of Steel Construction (AISC) Code, listed in Table 6.3.

This problem was previously solved in [1, 4, 6, 7]. The best results obtained using these approaches along with PC methodology for Case 1 and Case 2 are listed in Tables 6.4 and 6.5, respectively. The PC solution to the problem produced competent results at reasonable computational cost. The best, mean and worst function values, i.e. the weight W of the truss structure for Case 1 obtained from twenty trials were 477.16684 lb with zero standard deviation and for Case 2 were 464.14708 lb, 475.8719057 lb and 477.15846 lb, respectively with standard

Table 6.4 Performance comparison of various algorithms solving 25-bar case 1 truss structure problem

Variables	GA [1]	HS [4]	PSO [6]	PSOPC [6]	HPSO [6]	DHPSACO [7]	PC
x_1	0.4	0.01	0.01	0.01	0.01	0.01	0.01
$x_2 \sim x_5$	2.0	2.0	2.6	2.0	2.0	1.6	0.4
$x_6 \sim x_9$	3.6	3.6	3.6	3.6	3.6	3.2	3.6
$x_{10} \sim x_{11}$	0.01	0.01	00.01	0.01	0.01	0.01	0.01
$x_{12} \sim x_{13}$	0.01	0.01	0.4	0.01	0.01	0.01	2
$x_{14} \sim x_{17}$	0.8	0.8	0.8	0.8	0.8	0.8	0.8
$x_{18} \sim x_{21}$	2.0	1.6	1.6	1.6	1.6	2.0	0.01
$x_{22} \sim x_{25}$	2.4	2.4	2.4	2.4	2.4	2.4	4
$f(lb)$	563.52	560.59	566.44	560.59	560.59	551.61	477.1664

Table 6.5 Performance comparison of various algorithms solving 25-bar case 2 truss structure problem

Variables	GA [1]	PSO [6]	PSOPC [6]	HPSO [6]	DHPSACO [7]	Proposed PC
x_1	0.307	1.000	0.111	0.111	0.111	0.111
$x_2 \sim x_5$	1.990	2.620	1.563	2.130	2.130	0.563
$x_6 \sim x_9$	3.130	2.620	3.380	3.380	3.380	3.13
$x_{10} \sim x_{11}$	0.111	0.250	0.111	0.111	0.111	0.141
$x_{12} \sim x_{13}$	0.141	0.307	0.111	0.111	0.111	1.8
$x_{14} \sim x_{17}$	0.766	0.602	0.766	0.766	0.766	0.766
$x_{18} \sim x_{21}$	1.620	1.457	1.990	1.620	1.620	0.111
$x_{22} \sim x_{25}$	2.620	2.880	2.380	2.620	2.620	3.88
$f(lb)$	556.49	567.49	567.49	551.14	551.14	464.14708

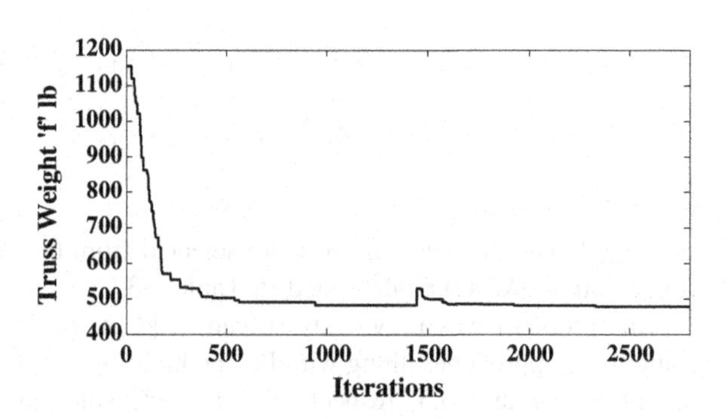

Fig. 6.4 Convergence plot for minimum weight of 25-bar case 1 truss structure problem

deviation 2.933655244. The average CPU time, average number of function evaluations and associated parameters are listed in Table 6.21. The convergence plots for Case 1 and Case 2 are shown in Figs. 6.4 and 6.5, respectively.

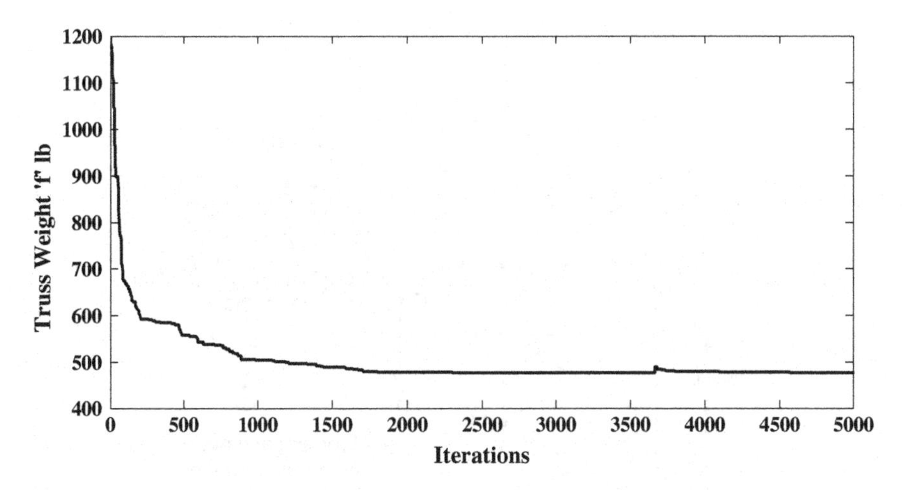

Fig. 6.5 Convergence plot for minimum weight of 25-bar case 2 truss structure problem

Problem 6.3: 72-Bar Truss Structure

For the 72-bar spatial truss structure shown in Fig. 6.6 was previously solve in [1, 4, 6, 7] has an aim to minimize the weight f (or W).

$$\text{Minimize } f = W = \sum_{i=1}^{N} \rho A_i l_i \tag{6.5}$$

$$\text{Subject to } |\sigma_i| \leq \sigma_{\max} \quad i = 1, 2, 3, \ldots, N \tag{6.6}$$

$$|u_j| \leq u_{\max} \quad j = 1, 2, 3, \ldots, M$$

The material density ρ is 0.1 lb/in^3 and the modulus of elasticity E is 10,000 ksi. The members are subjected to the stress limits $\sigma_{\max}$ of ± 25 ksi. The structure is subjected to load listed in Table 6.6. The nodes are subjected to the displacement limits $u_{\max}$ of 0.25 in. The 72 structural members of this spatial truss are sorted into 16 groups using symmetry: (1) $x_1 \sim x_4$, (2) $x_5 \sim x_{12}$ (3) $x_{13} \sim x_{16}$, (4) $x_{17} \sim x_{18}$, (5) $x_{19} \sim x_{22}$, (6) $x_{23} \sim x_{30}$, (7) $x_{31} \sim x_{34}$, (8) $x_{35} \sim x_{36}$, (9) $x_{37} \sim x_{40}$, (10) $x_{41} \sim x_{48}$, (11) $x_{49} \sim x_{52}$, (12) $x_{53} \sim x_{54}$, (13) $x_{55} \sim x_{58}$, (14) $x_{59} \sim x_{66}$, (15) $x_{67} \sim x_{70}$ and (16) $x_{71} \sim x_{72}$.

Two optimization cases are implemented.

For Case 1: The cross-section area of design variable for respective case should be selected from the set X_i = {0.1, 0.2, 0.3, 0.4, 0.5, 0.6, 0.7, 0.8, 0.9, 1.0, 1.1, 1.2, 1.3, 1.4, 1.5, 1.6, 1.7, 1.8, 1.9, 2.0, 2.1, 2.2, 2.3, 2.4, 2.5, 2.6, 2.7, 2.8, 2.9, 3.0, 3.1, 3.2} in^2.

For Case 2: The cross-section area of design variable for respective case should be selected from the Table 6.3.

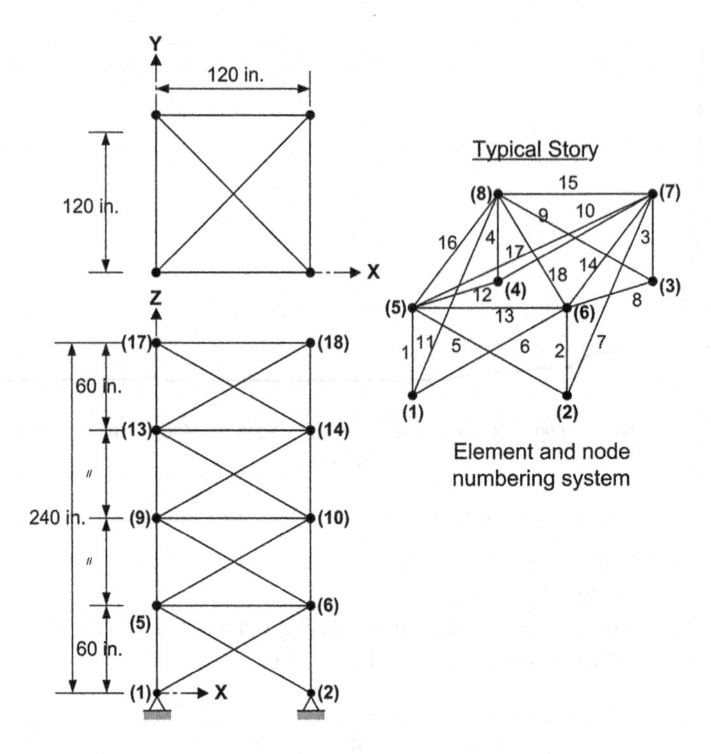

Fig. 6.6 72-Bar truss structure

Table 6.6 Loading condition for 72 bar

Nodes	P_x	P_y *kips*	P_z	P_x	P_y	P_z *kips*
17	5.0	5.0	0.0	0.0	0.0	−5.0
18	0.0	0.0	0.0	0.0	0.0	−5.0
19	0.0	0.0	0.0	0.0	0.0	−5.0
20	0.0	0.0	0.0	0.0	0.0	−5.0

This problem was previously solved in [1, 4, 6, 7]. The best results obtained using these approaches along with PC methodology for Case 1 and Case 2 are listed in Tables 6.7 and 6.8, respectively. The PC solution to the problem produced competent results at reasonable computational cost. The best, mean and worst function values, i.e. the weight W of the truss structure for Case 1 obtained from twenty trials were 372.40954 lb, 380.10692 lb and 395.99776 lb with standard deviation of 6.757504 and for Case 2 were 379.907983 lb, 382.329656 lb and 383.857921 lb, respectively with standard deviation 1.369460465. The average CPU time, average number of function evaluations and associated parameters are listed in Table 6.21. The convergence plots for Case 1 and Case 2 are shown in Figs. 6.7 and 6.8, respectively.

Table 6.7 Performance comparison of various algorithms solving 72-bar case 1 truss structure problem

Variables	GA [1]	HS [4]	PSO [6]	PSOPC [6]	HPSO [6]	DHPSA-CO [7]	Proposed PC
$x_1 \sim x_4$	1.5	1.9	2.6	3.0	2.1	1.9	2.1
$x_5 \sim x_{12}$	0.7	0.5	1.5	1.4	0.6	0.5	0.5
$x_{13} \sim x_{16}$	0.1	0.1	0.3	0.2	0.1	0.1	0.1
$x_{17} \sim x_{18}$	0.1	0.1	0.1	0.1	0.1	0.1	0.1
$x_{19} \sim x_{22}$	1.3	1.4	2.1	2.7	1.4	1.3	1.2
$x_{23} \sim x_{30}$	0.5	0.6	1.5	1.9	0.5	0.5	0.5
$x_{31} \sim x_{34}$	0.2	0.1	0.6	0.7	0.1	0.1	0.1
$x_{35} \sim x_{36}$	0.1	0.1	0.3	0.8	0.1	0.1	0.5
$x_{37} \sim x_{40}$	0.5	0.6	2.2	1.4	0.5	0.6	0.5
$x_{41} \sim x_{48}$	0.5	0.5	1.9	1.2	0.5	0.5	0.1
$x_{49} \sim x_{52}$	0.1	0.1	0.2	0.8	0.1	0.1	0.1
$x_{53} \sim x_{54}$	0.2	0.1	0.9	0.1	0.1	0.1	0.1
$x_{55} \sim x_{58}$	0.2	0.2	0.4	0.4	0.2	0.2	0.5
$x_{59} \sim x_{66}$	0.5	0.5	1.9	1.9	0.5	0.6	0.5
$x_{67} \sim x_{70}$	0.5	0.4	0.7	0.9	0.3	0.4	0.4
$x_{71} \sim x_{72}$	0.7	0.6	1.6	1.3	0.7	0.6	0.6
$f(lb)$	400.66	387.94	1089.88	1069.79	388.94	385.54	372.4095

Table 6.8 Performance comparison of various algorithms solving 72-bar case 2 truss structure problem

Variables	GA [1]	PSO [6]	PSOPC [6]	HPSO [6]	DHPSACO [7]	Proposed PC
$x_1 \sim x_4$	0.196	7.22	4.490	4.970	1.800	1.8
$x_5 \sim x_{12}$	0.602	1.80	1.457	1.228	0.442	0.563
$x_{13} \sim x_{16}$	0.307	1.13	0.111	0.111	0.141	0.111
$x_{17} \sim x_{18}$	0.766	0.196	0.111	0.111	0.111	0.141
$x_{19} \sim x_{22}$	0.391	3.09	2.620	2.880	1.228	1.457
$x_{23} \sim x_{30}$	0.391	0.785	1.130	1.457	0.563	0.443
$x_{31} \sim x_{34}$	0.141	0.563	0.196	0.141	0.111	0.111
$x_{35} \sim x_{36}$	0.111	0.785	0.111	0.111	0.111	0.111
$x_{37} \sim x_{40}$	1.800	3.090	1.266	1.563	0.563	0.602
$x_{41} \sim x_{48}$	0.602	1.228	1.457	1.228	0.111	0.443
$x_{49} \sim x_{52}$	0.141	0.111	0.111	0.111	0.250	0.111
$x_{53} \sim x_{54}$	0.307	0.563	0.111	0.196	0.196	0.111
$x_{55} \sim x_{58}$	1.563	0.990	0.442	0.391	0.563	0.111
$x_{59} \sim x_{66}$	0.766	1.620	1.457	1.457	0.442	0.563
$x_{67} \sim x_{70}$	0.141	1.563	1.228	0.766	0.766	0.443
$x_{71} \sim x_{72}$	0.111	1.266	1.457	1.563	1.563	0.563
$f(lb)$	427.203	1209.48	941.82	933.09	393.380	379.90798

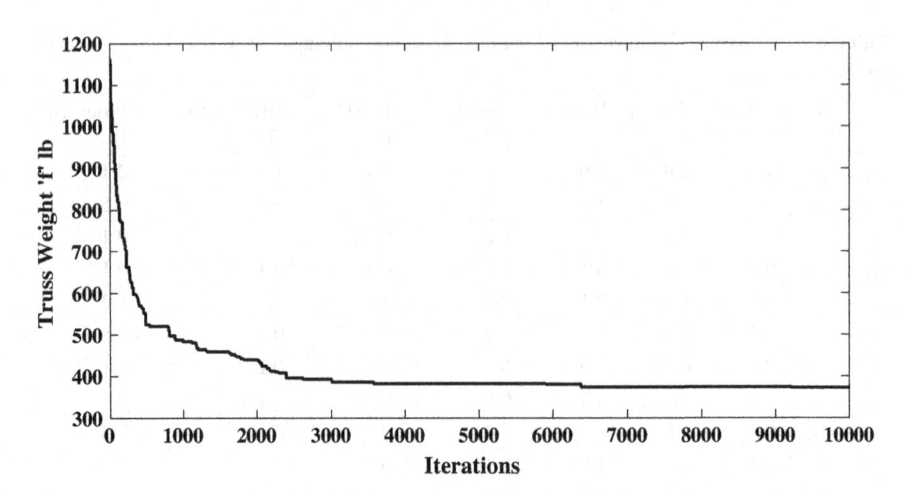

Fig. 6.7 Convergence plot for minimum weight of 72-bar case 1 truss structure problem

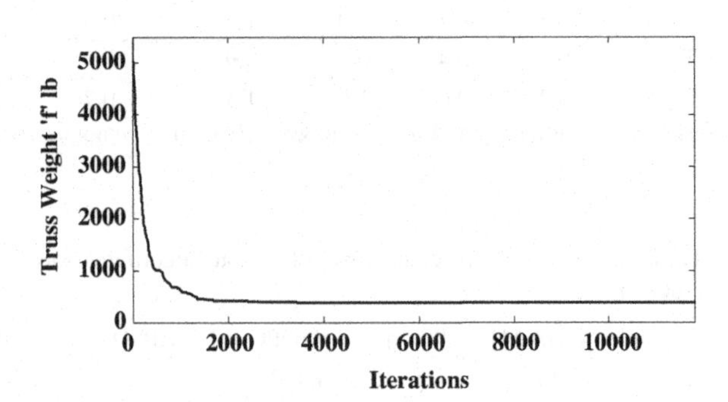

Fig. 6.8 Convergence plot for minimum weight of 72-bar case 2 truss structure problem

Problem 6.4: 45-Bar Truss Structure

This discrete variable problem discussed in [29, 31] aims to minimize the weight f (or W) of an AISI 1005 Steel 45-bar truss structure shown in Fig. 6.9. The problem formulation is as follows:

$$\text{Minimize}\, f = W = \sum_{i=1}^{N} \rho A_i l_i \tag{6.7}$$

$$\text{Subject to}\, |\sigma_i| \le \sigma_{\max} \quad i = 1, 2, 3, \ldots, N \tag{6.8}$$

$$|u_j| \le u_{\max} \quad j = 1, 2, 3, \ldots, M$$

where $N = 45$, $M = 20$, $\sigma_{\max} = 30$ ksi, $u_{\max} = \pm 2$ in, $\rho = 0.283\,\text{lb}/\text{in}^3$, E = 30,000 ksi and l_i is length of individual truss member i (refer to Fig. 6.9).

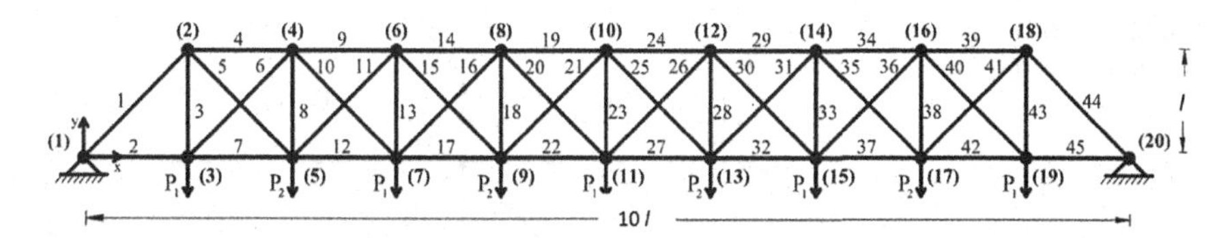

Fig. 6.9 45-Bar truss structure, $l = 200$ in

The truss members are linked according to the symmetry of the structure (refer to Fig. 6.9) into 23 groups considered as 23 sizing variables (refer to Table 6.9). The cross-section area of every member i, $(i = 1, 2, 3, \ldots, N)$ should be selected from the discrete set of $A_i = \{0.1, 0.2, \ldots, 14.9, 15\}$.

This problem was solved in International Student Competition in Structural Optimization (ISCSO) [29] and PC [31] as a discrete problem. In solving this problem using PC approach, truss members were considered as autonomous agents. Every truss member/agent i, $(i = 1, 2, 3, \ldots, N)$ was assigned cross-section area A_i as the strategy set.

The average CPU time, average number of function evaluations and associated parameters are listed in Table 6.21. The PC solution to the problem produced competent results with higher computational cost. The best, mean and worst function (i.e. the weight f or W) values obtained from 20 runs were 14377.9183 lb, 14554.8200 lb and 14761.00 lb, respectively, with standard deviation of 148.54. The convergence plot is shown in Fig. 6.16a.

Table 6.9 Performance comparison of various algorithms solving 45-bar truss structure problem

Variables in^2	ISCSO [29]	Proposed PC	Variables in^2	ISCSO [29]	Proposed PC
A_1, A_{44}	9.1000	9.7000	A_{13}, A_{33}	0.1000	0.1000
A_2, A_{45}	6.8000	7.0000	A_{14}, A_{29}	15.000	15.000
A_3, A_{43}	4.6000	5.0000	A_{15}, A_{31}	1.8000	1.9000
A_4, A_{39}	8.2000	7.9000	A_{16}, A_{30}	3.2000	2.9000
A_5, A_{41}	2.5000	2.3000	A_{17}, A_{32}	6.0000	5.6000
A_6, A_{40}	5.2000	5.5000	A_{18}, A_{28}	0.1000	0.1000
A_7, A_{42}	3.1000	3.3000	A_{19}, A_{24}	15.000	15.000
A_8, A_{38}	0.1000	0.3000	A_{20}, A_{26}	2.9000	3.0000
A_9, A_{34}	15.000	14.900	A_{21}, A_{25}	0.1000	0.1000
A_{10}, A_{36}	5.1000	4.8000	A_{22}, A_{27}	7.6000	7.1000
A_{11}, A_{35}	1.7000	1.8000	A_{23}	0.6000	0.8000
A_{12}, A_{37}	0.1000	0.1000			
Truss weight $f(lb)$				14341.21	14377.92

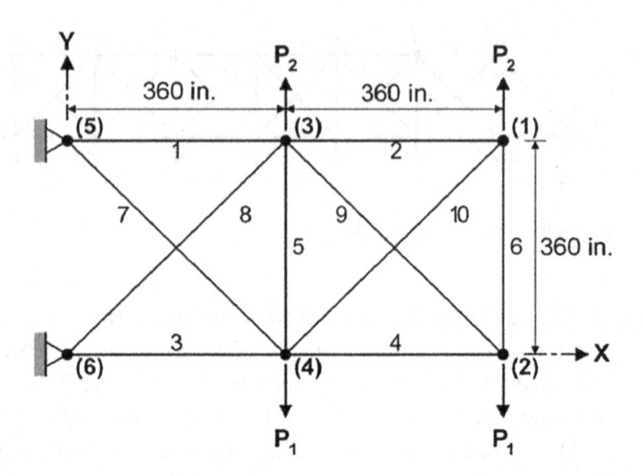

Fig. 6.10 10-Bar truss structure, $l = 360$ in

Problem 6.5: 10-Bar Truss Structure

This discrete variable problem discussed in [1, 6, 9, 17, 19] aims to minimize the weight f (or W) of an Aluminum 2024-T3 10-bar truss structure shown on Fig. 6.10. The problem formulation is as follows:

$$\text{Minimize}\, f = W = \sum_{i=1}^{N} \rho A_i l_i \tag{6.9}$$

$$\text{Subject to}\, |\sigma_i| \le \sigma_{\max} \quad i = 1, 2, 3, \ldots, N \tag{6.10}$$

$$|u_j| \le u_{\max} \quad j = 1, 2, 3, \ldots, M$$

where $N = 10$, $M = 6$, $\sigma_{\max} = \pm 25$ ksi, $u_{\max} = \pm 2$ in, $\rho = 0.1\ \text{lb/in}^3$, $E = 10{,}000$ ksi and l_i is length of individual truss member i (refer to Fig. 6.10).

Two distinct load cases were solved. For Case 1, the load P_1 and P_2 were chosen to be 100 kips and 0 (zero), respectively. The cross-section area A_i, $i = 1, 2, 3, \ldots, N$ was selected from the set $A_i = \{$1.62, 1.80, 1.99, 2.13, 2.38, 2.62, 2.63, 2.88, 2.93, 3.09, 3.13, 3.38, 3.47, 3.55, 3.63, 3.84, 3.87, 3.88, 4.18, 4.22, 4.49, 4.59, 4.80, 4.97, 5.12, 5.74, 7.22, 7.97, 11.50, 13.50, 13.90, 14.20, 15.50, 16.00, 16.90, 18.80, 19.90, 22.00, 22.90, 26.50, 30.00, 33.50$\}\text{in}^2$. For Case 2, the load P_1 and P_2 were chosen to be 150 kips and 150 kips, respectively. The cross-section area A_i, $i = 1, 2, 3, \ldots, N$ was selected from the set $A_i = \{$0.1, 0.5, 1.0, 1.5, 2.0, 2.5, 3.0, 3.5, 4.0, 4.5, 5.0, 5.5, 6.0, 6.5, 7.0, 7.5, 8.0, 8.5, 9.0, 9.5, 10.0, 10.5, 11.0, 11.5, 12.0, 12.5, 13.0, 13.5, 14.0,14.5, 15.0,15.5,16.0,16.5, 17.0, 17.5, 18.0, 18.5, 19.0, 19.5, 20.0, 20.5, 21.0, 21.5, 22.0, 22.5, 23.0, 23.5, 24.0, 24.5, 25.0, 25.5, 26.0, 26.5, 27.0, 27.5, 28.0, 28.5, 29.0, 29.5, 30.0, 30.5, 31.0, 31.5$\}\text{in}^2$.

The best results obtained using the approaches [1, 6, 9, 17, 19] along with the PC approach for solving Case 1 and Case 2 are presented in Tables 6.10 and 6.11, respectively.

Table 6.10 Performance comparison of various algorithms solving case 1 of the 10-bar truss structure problem

Variables in^2	SGA [1]	GA [9]	PSO [17]	PSOPC [6]	HPSO [6]	MBA [19]	Proposed PC
A_1	26.50	33.50	30.00	30.00	30.00	30.00	33.50
A_2	1.62	1.62	1.62	1.62	1.62	1.62	1.62
A_3	16	22	30	26.50	22.90	22.90	22.90
A_4	14.20	15.50	13.50	15.50	13.50	16.90	14.20
A_5	1.80	1.62	1.62	1.62	1.62	1.62	1.62
A_6	1.62	1.62	1.80	1.62	1.62	1.62	1.62
A_7	5.12	14.20	11.50	11.50	7.97	7.97	7.97
A_8	16.00	19.90	18.80	18.80	26.50	22.90	22.90
A_9	18.80	19.90	22.00	22.00	22.00	22.90	22.00
A_{10}	2.38	2.62	1.80	3.09	1.80	1.62	1.62
Truss weight $f(lb)$	4376.2	5613.84	5581.76	5593.44	5531.98	5507.75	5490.73789

Table 6.11 Performance comparison of various algorithms solving case 2 of the 10-bar truss structure problem

Variables in^2	SGA [1]	B&B [17]	PSO [6]	PSOPC [6]	HPSO [6]	MBA [19]	Proposed PC
A_1	30.50	30.50	24.50	31.50	31.50	29.50	23.50
A_2	0.50	0.10	0.10	0.10	0.10	0.01	0.10
A_3	16.50	23.00	22.50	23.50	24.50	24.00	26.00
A_4	15.00	15.50	15.50	18.50	15.50	15.00	14.00
A_5	0.10	0.10	0.10	0.10	0.50	0.01	0.10
A_6	0.10	0.50	1.50	0.50	0.50	0.05	2.00
A_7	0.50	7.50	8.50	7.50	7.50	7.50	12.50
A_8	18.00	21.00	21.50	21.50	20.50	21.50	13.00
A_9	19.50	21.50	27.50	23.50	20.50	21.50	20.00
A_{10}	0.50	0.10	0.10	0.10	0.10	0.10	0.10
Truss weight f(lb)	4217.30	5059.90	5243.71	5133.16	5073.51	5067.33	4686.77298

Every truss member i, $(i = 1, 2, 3, \ldots, N)$ was considered as autonomous agent selecting the strategy from within its associated strategy set A_i. The best, mean and worst function values (i.e. weight f or W) of the truss structure for Case 1 obtained from 20 runs were 5490.74 lb, 5491.81 lb and 5499.35 lb, respectively with standard deviation 2.67209 and for Case 2 were 4686.77298 lb, 4690.39948 lb and 4698.59636 lb, respectively with standard deviation 3.8299. The average CPU time, average number of function evaluations and associated parameters are listed in Table 6.21. The convergence plots for Case 1 and Case 2 are shown in Fig. 6.16b, c, respectively. The SGA method [1] provided the significantly better solution as it was incorporated with a two stage mapping process between decimal integers and the discrete values which increased diversity as well as local searching ability of the algorithm. In addition, this also resulted into reduced the computational efforts and cost.

Problem 6.6: 38-Bar Truss Structure

This discrete variable problem discussed in [30] aims to minimize the weight f (or W) of an AISI 1005 Steel 38-bar truss structure shown in Fig. 6.11. The problem formulation is as follows:

$$\text{Minimize } f = W = \sum_{i=1}^{N} \rho A_i l_i \tag{6.11}$$

$$\text{Subject to } |\sigma_i| \leq \sigma_{\max} \quad i = 1, 2, 3, \ldots, N \tag{6.12}$$

$$|u_j| \leq u_{\max} \quad j = 1, 2, 3, \ldots, M$$

where $P = 15$ kips, $N = 38$, $M = 21$, $\sigma_{\max} = 30$ ksi, $u_{\max} = \pm 4$ in, $\rho = 0.283\ \text{lb}/\text{in}^3$, $E = 30{,}000$ ksi and l_i is length of individual truss member i (refer to Fig. 6.11). The cross-section area of every member i, $(i = 1, 2, 3, \ldots, N)$ should be selected from the discrete set of $A_i = \{0.1, 0.2, 0.3, \ldots, 14.8, 14.9, 15\}$.

In solving this problem using PC approach, truss members were considered as autonomous agents. Every truss member/agent i, $(i = 1, 2, 3, \ldots, N)$ was assigned cross-section area A_i as the strategy set. And the best solutions obtained are presented in Table 6.12 in comparison with the solution [30]. The best, mean and worst PC solutions (i.e. weight f or W) from 20 runs were 5893.0227 lb, 5893.6567 lb and 5894.1950 lb, respectively, with standard deviation of 0.36962. The computational time, number of function evaluations and the associated parameters are listed in Table 6.21. The convergence plot is shown in Fig. 6.16d.

Problem 6.7: Helical Compression Spring Design

This mixed variable problem of helical compression spring design of Alloyed Steel [8, 10, 11, 18, 20, 22] presented in Fig. 6.12 aims to minimize the volume f (or V). The problem formulation is as follows:

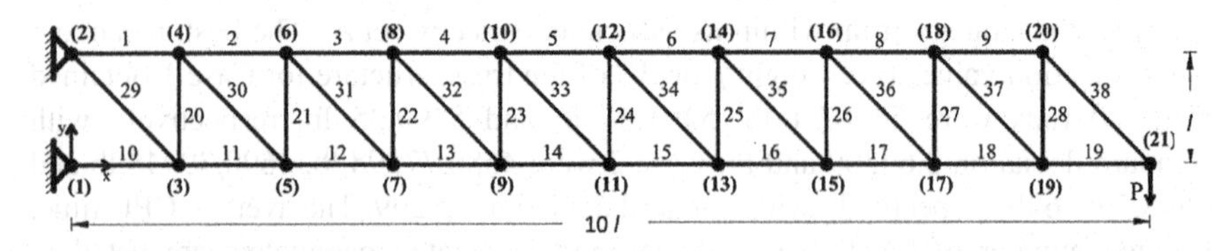

Fig. 6.11 38-Bar truss structure, $l = 100$ in

Table 6.12 Performance comparison of various algorithms solving 38-bar truss structure problem

Variables in^2	ISCSO [30]	Proposed PC	Variables in^2	ISCSO [30]	Proposed PC
A_1	14.60	14.60	A_{20}	1.60	1.60
A_2	12.90	12.90	A_{21}	1.60	1.60
A_3	11.30	11.50	A_{22}	1.60	1.70
A_4	9.70	9.60	A_{23}	1.60	1.60
A_5	8.20	7.80	A_{24}	1.60	1.60
A_6	6.50	6.50	A_{25}	1.60	1.80
A_7	4.90	4.90	A_{26}	1.60	1.60
A_8	3.30	3.10	A_{27}	1.60	1.60
A_9	1.70	1.60	A_{28}	1.60	1.60
A_{10}	15.00	150	A_{29}	2.30	2.30
A_{11}	14.60	14.80	A_{30}	2.30	2.50
A_{12}	12.90	12.80	A_{31}	2.30	2.30
A_{13}	11.30	11.30	A_{32}	2.30	2.40
A_{14}	9.70	9.70	A_{33}	2.30	2.30
A_{15}	8.20	8.10	A_{34}	2.30	2.30
A_{16}	6.50	6.30	A_{35}	2.30	2.30
A_{17}	4.90	4.70	A_{36}	2.30	2.40
A_{18}	3.30	3.50	A_{37}	2.30	2.40
A_{19}	1.70	1.60	A_{38}	2.30	2.30
Truss weight f(lb)				5889.9	5893.02273
No. function evaluations				4,618	1,793,966

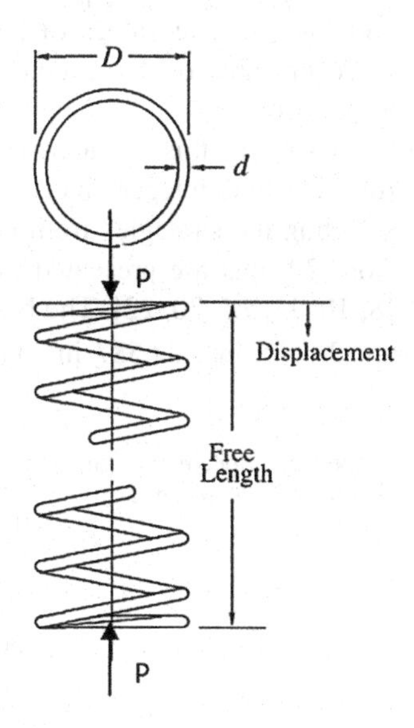

Fig. 6.12 Helical compression spring design problem

$$\text{Minimize } f = V = \frac{\pi^2 D d^2 (N+2)}{4} \tag{6.13}$$

Subject to

$$g_1 = \frac{8KP_{\max}D}{\pi d^3} - S \le 0 \tag{6.14}$$

$$g_2 = \left(\frac{P_{\max}}{k} - 1.05(N+2)d\right) - L_{free} \le 0 \tag{6.15}$$

$$g_3 = d_{\min} - d \le 0 \tag{6.16}$$

$$g_4 = (d + D) - D_{\max} \le 0 \tag{6.17}$$

$$g_5 = 3 - \frac{D}{d} \le 0 \tag{6.18}$$

$$g_6 = \delta_p - \delta_{pm} \le 0 \tag{6.19}$$

$$g_7 = \left(\frac{P_{\max}}{k} - 1.05(N+2)d - L_f\right) - L_{free} \le 0 \tag{6.20}$$

$$g_8 = \delta_w - \left(\frac{P_{\max} - P}{k}\right) \le 0 \tag{6.21}$$

where $C = \frac{D}{d}$, $K = \frac{4C-1}{4C-4} + \frac{0.615}{C}$, $k = \frac{Gd^4}{8ND^3}$, $\delta_p = \frac{P}{k}$, $1.0 \le D \le 30.0$, $N \ge 1$ and integer. The values of the discrete variable d can be taken from the values listed in Table 6.13 and constant terms require to solve the problem are listed in Table 6.14.

In solving this problem using PC approach, the variables D (continuous), N (discrete and integer) and d (discrete) were considered as autonomous agents selecting the strategies from their associated sets. And the best solutions obtained from 20 runs are presented in Table 6.15 in comparison with the approaches in [8, 10, 11, 18, 20, 22]. The best, mean and worst PC solutions (i.e. volume f or V) are 2.6585 in^3, 2.6587 in^3 and 2.6590 in^3, respectively, with standard deviation

Table 6.13 Specific design size for wire diameter d (in)

0.0090	0.0150	0.0280	0.0720	0.1620	0.2830
0.0095	0.0162	0.0320	0.0800	0.1770	0.3070
0.0104	0.0173	0.0350	0.0920	0.1920	0.3310
0.0118	0.0180	0.0410	0.1050	0.2070	0.3620
0.0128	0.0200	0.0470	0.1200	0.2250	0.3940
0.0132	0.0230	0.0540	0.1350	0.2440	0.4375

Table 6.14 The associated constant terms provided for the formulation of helical spring design problem

Constant terms	Description	Values
$P_{\max}$	Maximum work load	1000 lb
S	Maximum shear stress	$189 \times 10^3\, psi$
E	Elastic module of material	$30 \times 10^6\, psi$
G	Shear module of material	$11.5 \times 10^6\, psi$
L_{free}	Maximum coil free length	14 in
$d_{\min}$	Minimum wire diameter	0.2 in
$D_{\max}$	Maximum outside diameter of spring	3.0 in
P	Preload compression force	300.0 lb
δ_{pm}	Maximum deflection under preload	6.0 in
δ_w	Deflection from preload position to maximum load position	1.25 in

Table 6.15 Performance comparison of various algorithms solving helical spring design problem

Design Variables and Constraints	Nonlinear B&B [22]	HSIA [18]	MGA [10]	Gene-AS [8]	AHGA [11]	FA [20]	Proposed PC
d	0.2830	0.2830	0.2830	0.2830	0.2830	0.2830	0.2830
D	1.180701	1.223	1.227411	1.226	1.1096	1.223049	1.2231
N	10	9	9	9	9	9	9
g_1	−5430.9	−1008.81	−550.993	−713.51	25154.82	−1008.02	−1.0023
g_2	−8.8187	−8.946	−8.9264	−8.933	−9.1745	−8.946	−0.0089
g_3	−0.08298	−0.083	−0.0830	−0.0830	−0.0630	−0.083	−0.0001
g_4	−1.8193	−1.77696	−1.7726	−1.491	−1.890	−1.777	−0.0018
g_5	−1.1723	−1.3217	−1.3371	−1.337	−1.219	−1.322	−0.0055
g_6	−5.4643	−5.4643	−5.4485	−5.461	−5.464	−5.464	−0.0055
g_7	0	0	0	0	0	0	0
g_8	0	0.0001	−0.0134	−0.0090	−0.0014	0.0000	0
Spring volume $f(\text{in}^3)$	2.7995	2.659	2.6681	2.665	2.0283	2.6586	2.65860

0.0003. The computational time, number of function evaluations and the associated parameters are listed in Table 6.21. The standard deviation shows that the approach was sufficiently robust. The convergence plot is presented in Fig. 6.16e.

Problem 6.8: Reinforced Concrete Beam Design

This mixed variable problem discussed in [11, 20, 21, 23, 27] aims to minimize the total cost f of a reinforced cement concrete beam, shown on Fig. 6.13. The problem formulation is as follows:

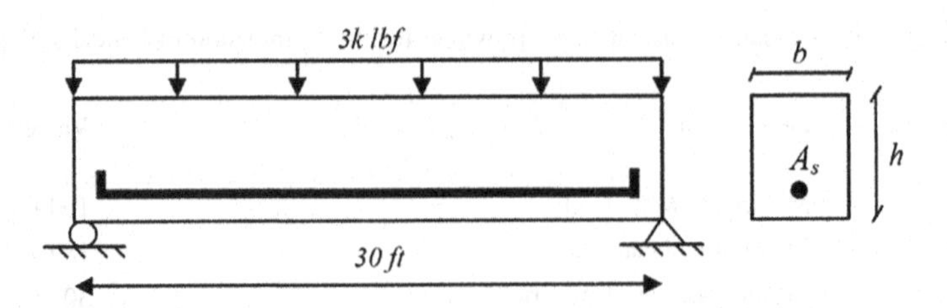

Fig. 6.13 Reinforced concrete beam

$$\text{Minimize}\, f = 29.4A + 0.6bh \tag{6.22}$$

$$\text{Subject to}\;\; g_1 = \frac{b}{h} - 4 \leq 0 \tag{6.23}$$

$$g_2 = 180 + 7.375\frac{A^2}{h} - Ab \leq 0 \tag{6.24}$$

$$M_u = 0.9A\sigma_y(0.8h)\left(1.0 - 5.59\frac{A\sigma_y}{0.8bh\sigma_c}\right) \geq 1.4M_d + 1.7M_l \tag{6.25}$$

where $\sigma_c = 5$ ksi, $\sigma_y = 500$ ksi, M_u = Flexural strength, $M_d = 1350$ kip.in, $M_l = 2700$ kip.in, $b = \{28, 29, \ldots, 39, 40\}$, $5 \leq h \leq 10$.

The value of A should be selected from the discrete set presented in Table 6.16.

The variables b and A are discrete, and variable h is continuous. In solving this problem using PC approach, these variables were considered as autonomous agents selecting the strategies from their associated sets. The best solutions obtained are presented in Table 6.17 in comparison with the approaches in [11, 20, 21, 23, 27]. The PC solution produced competent results with higher computational cost. The best, mean and worst PC solutions (i.e. cost f) obtained from 20 runs conducted were 359.2080, 359.5669 and 362.2500, respectively, with the standard deviation 1.0098. The computational time, number of function evaluation, the associated parameters are listed in Table 6.21. The convergence plot for the best solution is presented in Fig. 6.16f.

Problem 6.9: Stepped Cantilever Beam Design

This mixed variable problem discussed in [12–14, 16, 20, 26, 28] aims to minimize the volume f (or V) of the stepped cantilever beam of Alloyed Steel presented in Fig. 6.14. The problem formulation is as follows:

$$\text{Minimize}\; f = V = l(b_1h_1 + b_2h_2 + b_3h_3 + b_4h_4 + b_5h_5)a \tag{6.26}$$

Table 6.16 Discrete values of cross section area $A\ (\mathrm{in}^2)$

Bar type	A	Bar type	A	Bar type	A	Bar type	A
1#4	0.2	6#5	1.86	9#6	3.95	9#8	7.11
1#5	0.31	10#4, 2#9	2	4#9	3.96	12#7	7.2
2#4	0.4	7#5	2.17	13#5	4	13#7	7.8
1#6	0.44	11#4, 5#6	2.2	7#7	4.03	10#8	7.9
3#4, 1#7	0.6	3#8	2.37	14#5	4.2	8#9	8
2#5	0.62	12#4, 4#7	2.4	10#6	4.34	14#7	8.4
1#8	0.79	8#5	2.48	15#5	4.4	11#8	8.69
4#4	0.8	13#4	2.6	6#8	4.65	15#7	9
2#6	0.88	6#6	2.64	8#7	4.74	12#8	9.48
3#5	0.93	9#5	2.79	11#6	4.8	13#8	10.27
5#4, 1#9	1	14#4	2.8	5#9	4.84	11#9	11
6#4, 2#7	1.2	15#4, 5#7, 3#9	3	12#6	5	14#8	11.06
4#5	1.24	7#6	3.08	9#7	5.28	15#8	11.85
3#6	1.32	10#5	3.1	7#8	5.4	12#9	12
7#4	1.4	4#8	3.16	13#8	5.53	13#9	13
5#5	1.55	11#5	3.41	10#7, 6#9	5.72	14#9	14
2.8	1.58	8#6	3.52	14#6	6	15#9	15
8#4	1.6	6#7	3.6	8#8	6.16		
4#6	1.76	12#5	3.72	15#6, 11#7	6.32		
9#4, 3#7	1.8	5#8	3.82	7#9	6.6		

Subject to

$$g_1 = \frac{6Pl}{b_5 h_5^2} - \sigma_d \leq 0 \tag{6.27}$$

$$g_2 = \frac{6P(2l)}{b_4 h_4^2} - \sigma_d \leq 0 \tag{6.28}$$

$$g_3 = \frac{6P(3l)}{b_3 h_3^2} - \sigma_d \leq 0 \tag{6.29}$$

$$g_4 = \frac{6P(4l)}{b_2 h_2^2} - \sigma_d \leq 0 \tag{6.30}$$

$$g_5 = \frac{6P(5l)}{b_1 h_1^2} - \sigma_d \leq 0 \tag{6.31}$$

$$g_6 = \frac{Pl^3}{3E}\left(\frac{1}{I_5} + \frac{7}{I_4} + \frac{19}{I_3} + \frac{37}{I_2} + \frac{61}{I_1}\right) - D_{\max} \leq 0 \tag{6.32}$$

Table 6.17 Performance comparison of various algorithms solving reinforced concrete beam design problem

Design variables and constraints	SD-RC [23]	GHN-ALM GHN-EP [21]		GA GA-FL [11]		FA [20]	BFBA [27]	Proposed PC
A_i	7.8	6.6	6.32	7.20	6.16	6.32	N/A	6.32
b	31	33	34	32	35	34	N/A	34
h	7.79	8.495227	8.637180	8.0451	8.7500	8.50	N/A	8.5
g_1	−4.2012	0.0159	−0.7745	−2.8779	−3.6173	−0.2241	N/A	−0.2241
g_2	−0.0205	−0.1155	−0.0635	−0.0224	0	0	N/A	0
f	374.2	362.2455	362.00648	366.1459	364.8541	359.2080	376.2977	359.2080229
No. function evaluations	396	N/A	N/A	100,000	100,000	30,000	25,000	56,3490

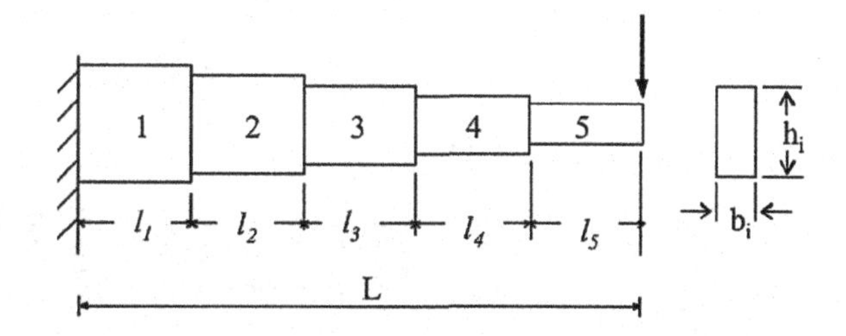

Fig. 6.14 Steeped cantilever beam

$$g_7 = \frac{h_5}{b_5} - 20 \leq 0 \tag{6.33}$$

$$g_8 = \frac{h_4}{b_4} - 20 \leq 0 \tag{6.34}$$

$$g_9 = \frac{h_3}{b_3} - 20 \leq 0 \tag{6.35}$$

$$g_{10} = \frac{h_2}{b_2} - 20 \leq 0 \tag{6.36}$$

$$g_{11} = \frac{h_1}{b_1} - 20 \leq 0 \tag{6.37}$$

where $b_1, b_2, b_3, h_1, h_2, h_3$ are discrete variables and b_4, b_5, h_4, h_5 are continuous variables. The corresponding sampling sets and intervals are defined as follows: $b_1 = \{1, 2, 3, 4, 5\}$, $b_2, b_3 = \{2.4, 2.6, 2.8, 3.1\}$, $h_1, h_2 = \{45, 50, 55, 60\}$, $h_3 = \{30, 31, \ldots, 64, 65\}$, $1 \leq b_4, b_5 \leq 5$ and $30 \leq h_4, h_5 \leq 65$. The values of constant terms are listed in Table 6.18.

In solving this problem using PC approach, the variables $b_1, b_2, b_3, h_1, h_2, h_3, h_4$ and h_5 were considered as autonomous agents selecting the strategies from their associated sets. And the best solutions obtained are presented in Table 6.19 in comparison with the approaches in [12–14, 16, 20, 26, 28]. The best, mean and worst PC solutions (i.e. volume f or V) obtained from 20 runs conducted were 61473.93, 63078.44 and 64420.69 cm^3, respectively, with standard deviation

Table 6.18 Constant terms provided for the formulation of stepped beam design problem

Constant terms	Description	Values
P	Concentrated load	1000 lb
σ_d	Design bending stress	189×10^3 *psi*
E	Elastic modulus of the material	30×10^6 *psi*
$D_{\max}$	Allowable deflection	11.5×10^6 *psi*
L	Total length of the five-stepped cantilever beam	14 in

Table 6.19 Performance comparison of various algorithms solving stepped beam design problem

Design variables	RNES [26]	B&B - RU [28]	B&B - PD [28]	B&B - LAD [28]	B&B - CAD [28]	GAOS [13]	GA-APM [12]	AIS-GA [14]	AIS-GA-C [14]	FA [20]	Proposed PC
b_1	3.0000	4.0000	3.0000	3.0000	3.0000	3.0000	3.0000	3.0000	3.0000	3.0000	3.0000
h_1	60.0000	62.0000	60.0000	60.0000	60.0000	60.0000	60.0000	60.0000	60.0000	60.0000	60.0000
b_2	3.1000	3.1000	3.1000	3.1000	3.1000	3.1000	3.1000	3.1000	3.1000	3.1000	3.1000
h_2	55.0000	60.0000	55.0000	55.0000	55.0000	55.0000	55.0000	55.0000	60.0000	55.0000	55.0000
b_3	2.6000	2.6000	2.6000	2.6000	2.6000	2.6000	2.6000	2.6000	2.6000	2.6000	2.6000
h_3	50.0000	55.0000	50.0000	50.0000	50.0000	50.0000	50.0000	50.0000	50.0000	50.0000	50.0000
b_4	2.3110	2.2050	2.2760	2.2620	2.2790	2.2700	2.2890	2.2350	2.3110	2.2050	2.2679
h_4	43.1080	44.0900	45.5280	45.2330	45.5530	45.2500	45.6260	44.3950	43.1680	44.0910	45.3540
b_5	1.8220	1.7510	1.7500	1.7500	1.7500	1.7500	1.7930	2.0040	2.2250	1.7500	3.6500
h_5	34.3070	35.0300	34.9950	34.9950	35.0040	35.0000	34.5930	32.8790	31.2500	34.9950	31.3420
Volume $f(\text{cm}^3)$	64269.59	73555.00	64537.00	64403.00	64558.00	64447.00	64698.56	65559.60	66533.47	63893.52	61473.93

1023.508. The computational time, number of function evaluations and the associated parameters are listed in Table 6.21. The convergence plot is presented in Fig. 6.10g.

Problem 6.10: Speed Reducer Design

This mixed variable problem is discussed in [14, 15, 24, 25] aims to minimize the weight f (or W) of the speed reducer of material 303 Stainless Steel presented in Fig. 6.15. The problem is formulated as follows:

$$\begin{aligned}\text{Minimize}\quad f = W = {} & 0.7854bz^2(3.3333Z^3 + 14.9334Z - 43.0934) \\ & - 1.508b\left(D_1^2 + D_2^2\right) + 7.4777\left(D_1^3 + D_2^3\right) \\ & + 0.7854\left(L_1D_1^2 + L_2D_2^2\right)\end{aligned} \tag{6.38}$$

Subject to

$$g_1 = \frac{27}{bz^2Z} - 1 \le 0 \tag{6.39}$$

$$g_2 = \frac{397.5}{bz^2Z^2} - 1 \le 0 \tag{6.40}$$

$$g_3 = \frac{1.93D_1}{zZL_1^4} - 1 \le 0 \tag{6.41}$$

$$g_4 = \frac{1.93D_2^3}{zZL_2^4} - 1 \le 0 \tag{6.42}$$

$$g_5 = \frac{\sqrt{\left(\frac{745D_1}{zZ}\right)^2 + 16.9e6}}{110L_1^3} - 1 \le 0 \tag{6.43}$$

$$g_6 = \frac{\sqrt{\left(\frac{745D_2}{zZ}\right)^2 + 157.5e6}}{85L_2^3} - 1 \le 0 \tag{6.44}$$

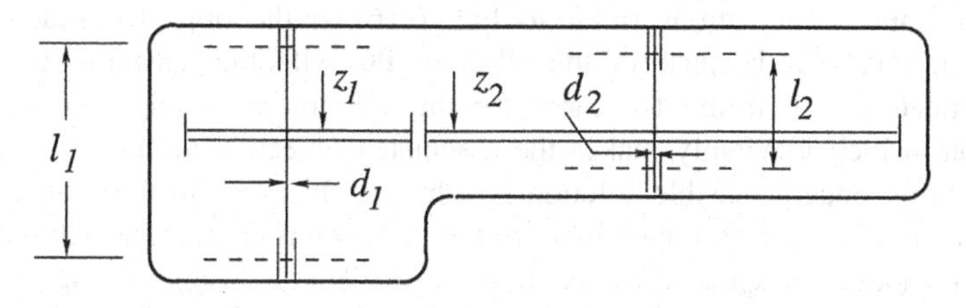

Fig. 6.15 Speed reducer

$$g_7 = \frac{zZ}{40} - 1 \le 0 \tag{6.45}$$

$$g_8 = \frac{5z}{b} - 1 \le 0 \tag{6.46}$$

$$g_9 = \frac{b}{12z} - 1 \le 0 \tag{6.47}$$

$$g_{10} = \frac{1.5L_1 + 1.9}{D_1} - 1 \le 0 \tag{6.48}$$

$$g_{11} = \frac{1.1L_2 + 1.9}{D_2} - 1 \le 0 \tag{6.49}$$

$2.6 \le b \le 3.6$, $0.7 \le z \le 0.8$, $7.3 \le D_1 \le 8.3$, $7.8 \le D_2 \le 8.3$, $2.9 \le L_1 \le 3.9$, $5 \le L_2 \le 5.5$, and $17 \le Z \le 28$ and integer.

The variables b, z, D_1, D_2, L_1, L_2 are continuous and Z is a discrete variable. In solving this problem using PC approach, these variables were considered as autonomous agents selecting the strategies from their associated sets/sampling intervals. The results obtained are presented in Table 6.20 in comparison with the approaches in [14, 15, 24, 25]. The PC solution to the problem produced competent results with higher computational cost. The best, mean and worst PC solutions (i.e. weight f or W) obtained from 20 runs conducted were 2828.5863 lb, 2854.91 lb and 2855.62 lb, respectively with standard deviation 6.60299. The computational time, number of function evaluations and the associated parameters are listed in Table 6.21. The convergence plot is presented in Fig. 6.16h.

6.1 Discussion and Conclusions

The PC approach was successfully applied for solving practically important discrete and mixed variable problems from structural as well as mechanical engineering design domain. In addition, the feasibility-based rule is also validated as a powerful tool for handling a variety of constraints. In majority of the problems solved here initial variations/fluctuations (refer to Fig. 6.16) in the objective function was observed. This was because of the effort of PC with the feasibility-based rule incorporated into it trying to search for the solution with improved constraint violations which inevitably makes the associated objective function to fluctuate. Furthermore, once a feasible solution is achieved, in every further iteration only the feasible solution with improved objective function is accepted which makes the solution smoothly progress towards convergence. This is evident in Fig. 6.16. The

Table 6.20 Performance comparison of various algorithms solving weight optimization of speed reducer problem

Design variables and constraints	EA [24]	AIS-GA [15]	AIS-GA [14]	AIS-GA-C [14]	DEA [25]	Proposed PC
b	3.506163	3.500000	3.500001	3.500000	N.A.	3.506102231
z	0.700831	0.700000	0.700000	0.700000	N.A.	0.700173506
Z	17.0	17.0	17.0	17.0	N.A.	17
D_1	7.460181	7.300008	7.300017	7.3000035	N.A.	7.315691279
D_2	7.962143	7.715322	7.715326	7.7153225	N.A.	7.778379574
L_1	3.362900	3.350215	3.350216	3.3502147	N.A.	3.356601485
L_2	5.308949	5.286655	5.286654	5.2866545	N.A.	5.000832551
g_1	−0.077734	−0.0739152	−0.07391554	−0.07391524	N.A.	−0.075985213
g_2	−0.201305	−0.19799852	−0.19799876	−0.19799852	N.A.	−0.236429611
g_3	−0.474119	−0.49917084	−0.49916983	−0.49917156	N.A.	−0.499885213
g_4	−0.897068	−0.90464383	−0.90464365	−0.90464383	N.A.	−0.877988474
g_5	−0.011021	$-2.3 \times 10 - 7$	$-1.55 \times 10 - 6$	$-1.19 \times 10 - 7$	N.A.	−0.00390051
g_6	−0.012500	$-2.9 \times 10 - 7$	0.00	0.00	N.A.	−0.623858326
g_7	−0.702147	−0.70250000	−0.70250000	−0.70250000	N.A.	−0.70242626
g_8	−0.000573	0.00000000	$-2.9 \times 10 - 7$	0.00000000	N.A.	−0.001493026
g_9	−0.583095	−0.5833333	−0.58333320	−0.58333330	N.A.	−0.582710309
g_{10}	−0.583095	−0.5833333	−0.58333320	−0.58333330	N.A.	−0.052051001
g_{11}	−0.027920	−0.00000018	−0.00000077	−0.00000036	N.A.	−0.048527301
f	3025.0051	2994.3419	2994.4720	2994.4712	2863.36	2828.5863
No. function evaluations	36,000	150,000	36,000	36,000	N.A.	1,132,700

NA Not Available

Table 6.21 PC solution details

Problems	System sol. best mean worst	Standard deviation	Average number of function evaluations	Average computational time (min)	Closeness to the best reported solution %	Set of parameters $m_i\{\lambda_{up}$ and $\lambda_{down}\}, \Gamma, \alpha_T$
17-bar truss structure	2584.02556	1.3755	6,628,943	18.78	−0.0455	7, 0.001^c, 40, 0.05
	2586.11318					
	2589.25345					
25-bar truss structure Case 1	477.16684	0	1,844,457	2.36	13.495*	7, 1^d, 500, 0.005
	477.16684					
	477.16684					
Case 2	476.43010	0.2164	1,963,415	4.56	13.555*	7, 1^d, 500, 0.005
	476.56160					
	477.15846					
72-Bar truss structure						
Case 1	372.40954	6.7575	8,843,207	34.45	3.424*	7, 1^d, 500, 0.005
	380.10692					
	395.99776					
Case 2	379.90798	1.3694	8,730,598	30.99	3.405*	7, 1^d, 500, 0.005
	382.32966					
	383.85792					
45-bar truss structure	14377.9183	148.54	4,494,278	62.5	0.2559	7, $\{1^d$ and $1^d\}$, 450, 0.065
	14554.82					
	14761					

(continued)

Table 6.21 (continued)

Problems	System sol. best mean worst	Standard deviation	Average number of function evaluations	Average computational time (min)	Closeness to the best reported solution %	Set of parameters $m_i\{\lambda_{up}$ and $\lambda_{down}\}, \Gamma, \alpha_T$
10-bar truss structure						
Case: 1	5490.74	2.67209	1,852,059	13.03	25.46	7, $\{1^d$ and $1^d\}$, 870, 0.0076
	5491.81					
	5499.35					
Case: 2	4686.77298	3.8299	2,363,380	1.64	0.02	7, $\{1^d$ and $1^d\}$, 900, 0.005
	4690.39948					
	4698.56636					
38-Bar truss structure	5893.0227	0.36962	1,793,966	22.49	0.0514	7, $\{1^d$ and $1^d\}$, 760, 0.0057
	5893.6567					
	5894.1950					
Helical Spring	2.65857	0.00013	498,567	1.94	0.0001*	7, $\{2^i$ and $2^i\}$, 510, 0.007 $\{1^d$ and $1^d\}$ $\{0.1^d$ and $0.1^d\}$
	2.6587					
	2.6590					
Reinforced Concrete Beam	359.2080	1.0098	563,490	1.66	0.00	7, $\{1^d$ and $1^d\}$, 550, 0.005
	359.5669					
	362.2500					
Stepped Cantilever Beam	61473.93	1023.508	5,075,700	24.8	3.786*	7, $\{1^d$ and $1^d\}$, 800, 0.05 $\{0.1^c$ and $0.1^c\}$
	63078.44					
	64420.69					

(continued)

Table 6.21 (continued)

Problems	System sol. best mean worst	Standard deviation	Average number of function evaluations	Average computational time (min)	Closeness to the best reported solution %	Set of parameters $m_i\{\lambda_{up}$ and $\lambda_{down}\}, \Gamma, \alpha_T$
Speed Reducer	2828.5863	6.60299	1,055,840	2.598	1.214*	7, $\{1^i$ and $1^i\}$, 550, 0.005 $\{0.01^c$ and $0.01^c\}$
	2854.91					
	2855.62					

^{I} Integer variable
^{d} Discrete variable
^{c} Continuous variable
* Shows the optimal design obtained using PC was better than other algorithms

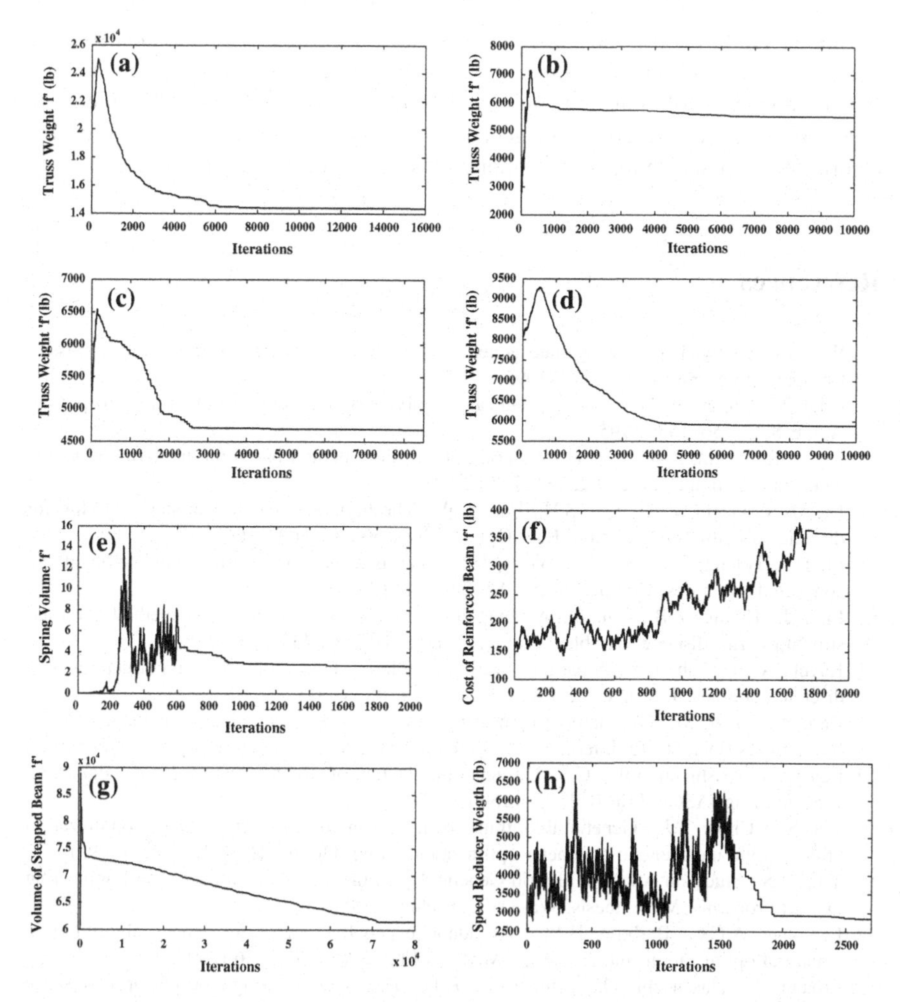

Fig. 6.16 Convergence plots for solved problems **a** convergence plot for 45-bar truss structure problem **b** convergence plot for case-1 of the 10-bar truss structure problem **c** convergence plot for case-2 of the 10-bar truss structure problem **d** convergence plot for 38-bar truss structure problem **e** convergence plot for helical spring problem **f** convergence plot for reinforced concrete beam problem **g** convergence plot for stepped beam problem **h** convergence plot for speed reducer problem

choice of the shrinking interval factor λ_{down} may help to expedite the convergence process as it allows the agents to choose the strategies from within the close neighborhood of their current favorable strategies and may also increase the chances to search better solution. In addition, the choice of expansion interval factor λ_{up} which expands the sampling interval plays an important role here. It gives a chance to every agent to jump out of the local minima and may further help to search for a better solution; however, it may also result in a wider sampling space for the agents to choose the strategies from within forming the combined strategy sets. This may result in larger fluctuations in the objective function when searching for the improved solutions (i.e. improved combined strategy set) in the infeasible space as well as delayed convergence and possibly reduced robustness of the algorithm when searching for the improved solution in the feasible space.

References

1. Wu, S.J., Chow, P.T.: Steady-state genetic algorithms for discrete optimization of trusses. Comput. Struct. **56**(6), 979–991 (1995)
2. Adeli, H., Kumar, S.: Distributed genetic algorithm for structural optimization. J. Aerosp. Eng. ASCE **8**(3), 156–163 (1995)
3. Lee, K.S., Geem, Z.W.: A new structural optimization method based on the harmony search algorithm. Comput. Struct. **82**, 781–798 (2004)
4. Lee, K.S., Geem, Z.W., Lee, S.H., Bae, K.W.: The harmony search heuristic algorithm for discrete structural optimization. Eng. Optim. **37**(7), 663–684 (2005)
5. Li, L.J., Huang, Z.B., Liu, F., Wu, Q.H.: A heuristic particle swarm optimization of pin connected structures. Comput. Struct. **85**, 340–349 (2007)
6. Li, L.J., Huang, Z.B., Liu, F.: A heuristic particle swarm optimization method for truss structures with discrete variables. Comput. Struct. **87**(7–8), 435–443 (2009)
7. Kaveh, A., Talatahari, S.: A particle swarm ant colony optimization for truss structures with discrete variables. J. Struct. Steel Res. **65**, 1558–1568 (2009)
8. Deb, K. Goyal, M.: Optimizing engineering designs using a combined genetic search. In: Proceedings Back IT, 7th International Conference of Genetic Algorithm, pp. 512–528 (1997)
9. Rajeev, S., Krishnamoorthy, C.S.: Discrete optimization of structures using genetic algorithm. J. Struct. Eng. ASCE **118**(5), 1123–1250 (1992)
10. Wu, S.J., Chow, P.E.: Genetic algorithms for nonlinear mixed discrete-integer optimization problems via metagenetic parameter optimizations. Eng. Optim. **24**(2), 137–159 (1995)
11. Yun, Y.S.: Study on Adaptive Hybrid Genetic Algorithm and its Applications to Engineering Design Problems. M.Sc. thesis, Waseda University (2005)
12. Lemonge, A.C.C., Barbosa, H.J.C.: An adaptive penalty scheme for genetic algorithms in structural optimization. Int. J. Numer. Methods Eng. **59**(5), 703–736 (2004)
13. Erbatur, F., Hasancebi, O., Tutuncu, I., Kilic, H.: Optimal design of planar and space structures with genetic algorithms. Comput. Struct. **75**(2), 209–224 (2000)
14. Bernardino, H.S., Barbosa, H.J.C., Lemonge, A.C.C.: A hybrid genetic algorithm for constrained optimization problems in mechanical engineering. In: Proceedings IEEE Congress Evolution Computations, pp. 646–653 (2007)
15. Coello, C.A.C., Cortes, N.C.: Hybridizing a genetic algorithm with an artificial immune system for global optimization. Eng. Optim. **36**(5), 607–634 (2004)
16. Azad, S.K., Azad, S.K., Kulkarni, A.J.: Structural optimization using a mutation-based genetic algorithm. Int. J. Optim. Civil Eng. **2**(1), 80–100 (2012)
17. Ringertz, U.T.: On methods for discrete structural constraints. Eng. Optim. **13**(1), 47–64 (1988)
18. Guo, C.X., Hu, J.S., Ye, B., Cao, Y.J.: Swarm intelligence for mixed-variable design optimization. J. Zhejiang Univ. Sci. **5**(7), 851–860 (2004)
19. Sadollah, A., Bahreininejad, A., Eskandar, H., Hamdi, M.: Mine blast algorithm for optimization of truss structures with discrete variables. Comput. Struct. **49**(63), 102–103 (2012)
20. Gandomi, A.H., Xin-She Yang., Alavi, A.H.: Mixed variable structural optimization using Firefly algorithm. Comput. Struct. 89(23–24), pp. 2325–2336 (2011)

21. Shih, C.J., Yang, Y.C.: Generalized Hopfield network based structural optimization using sequential unconstrained minimization technique with additional penalty strategy, Adv. Eng. Softw. 33 (7–10), pp. 721–729 (2002)
22. Sandgren, E.: Nonlinear integer and discrete programming in mechanical design optimization. J. Mech. Des. **112**(2), 223–229 (1990)
23. Amir, H.M., Hasegawa, T.: Nonlinear mixed-discrete structural optimization. J. Struct. Eng. **115**(3), 626–645 (1989)
24. Efren, M., Coello, C.A.C., Ricardo, L.: Engineering optimization using a simple evolutionary algorithm. In: Proceedings 15th International Conference on Tools with Artificial Intelligence (ICTAI), pp. 149–156 (2003)
25. Pant, M., Thangaraj, R., Singh, V.P.: Optimization of mechanical engineering design problems using improved differential evolutionary algorithm. Int. J. Resent Trend Eng. **1**(5), 21–25 (2009)
26. Chen, T.Y., Chen, H.C.: Mixed-discrete structural optimization using a rank-niche evolution strategy. Eng. Optim. **41**(1), 39–58 (2009)
27. Montes, E.M., Ocana, B.H.: Modified bacterial foraging optimization for engineering design. In: C.H. Dagli, et al., of the Artificial Neural Networks in Engineering Conference (ANNIE), ASME Press series, Intelligent Engineering Systems through Artificial Neural Networks, 19, pp. 357–364 (2009)
28. Thanedar, P.B., Vanderplaats, G.N.: Survey of discrete variable optimization for structural design. J. Struct. Eng. **121**(2), 301–306 (1995)
29. Arnout, S.: International Student Competition in Structural Optimization, Aug. 26–27 (2011)
30. Rudolph, S., Schmidt, J.: International Student Competition in Structural Optimization, Aug. 26–27 (2012)
31. Kulkarni, A.J., Kale, I.R., Tai, K.: Probability collectives for solving discrete and mixed variable problems. In: International Journal of Computer Aided Engineering and Technology (2014) (In Press)
32. Kulkarni, A.J., Kale, I.R., Tai, K., Azad, S.K.: Discrete optimization of truss structure using probability collectives. In: Proceedings IEEE 12th International Conference of Hybrid Intelligence System, pp. 225–230 (2002)
33. Kulkarni, A.J., Kale, I.R., Tai, K.: Probability collectives for solving truss structure problems. In: 10th World Congress on Structural and Multidisciplinary Optimization, paper no. 5395 (2013)

Chapter 7
Probability Collectives with Feasibility-Based Rule II

This rule is a variation of the Feasibility-based Rule I discussed in Chap. 5 and also allows the objective function and the constraint information to be considered separately. In addition to the iterative tightening of the constraint violation tolerance in order to obtain fitter solution and further drive the convergence towards feasibility, the Feasibility based Rule II helps the solution jump out of possible local minima.

Consider a general constrained problem (in the minimization sense) as follows:

$$\begin{aligned} &\text{Minimize} \quad G \\ &\text{Subject to} \quad g_j \le 0\,, \quad j = 1, 2, \ldots, s \\ &\qquad\qquad\quad h_j = 0\,, \quad j = 1, 2, \ldots, w \end{aligned} \tag{7.1}$$

According to [1–3], the equality constraint $h_j = 0$ can be transformed into a pair of inequality constraints using a tolerance value δ as follows:

$$h_j = 0 \quad \Rightarrow \begin{cases} g_{s+j} = h_j - \delta \le 0 & j = 1, 2, \ldots, w \\ g_{s+w+j} = -\delta - h_j \le 0 \end{cases} \tag{7.2}$$

Thus, w equality constraints are replaced by $2w$ inequality constraints with the total number of constraints given by $t = s + 2w$. Then a generalized representation of the problem in Eq. (7.1) can be stated as follows:

$$\begin{aligned} &\text{Minimize} \quad G \\ &\text{Subject to} \quad g_j \le 0\,, \quad j = 1, 2, \ldots, t \end{aligned} \tag{7.3}$$

The following sections discuss the Feasibility-based Rule II, associated formulation, implementation and validation of results solving three cases of the Sensor Network Coverage Problem (SNCP) as well as several cases of the well known constrained and unconstrained test problems.

© Springer International Publishing Switzerland 2015

A.J. Kulkarni et al., *Probability Collectives*, Intelligent Systems Reference Library 86, DOI 10.1007/978-3-319-16000-9_7

7.1 Feasibility Based Rule II

At the beginning of the PC algorithm, the number of constraints improved μ is initialized to zero, i.e. $\mu = 0$. The value of μ is updated iteratively as the algorithm progresses. Similar to the Feasibility-based Rule I, Feasibility-based Rule II is also assisted with the modified approach of updating of the sampling space as well as the convergence criterion. The rule and these assisting techniques are discussed below as the modifications to the unconstrained PC approach presented in Sect. 2.2.1.

7.1.1 Modifications to Step 5 of the Unconstrained PC Approach

The step 5 of the unconstrained PC procedure discussed in Sect. 2.2.1 is modified by using the following rule:

Any feasible solution is preferred over any infeasible solution
Between two feasible solutions, the one with better objective is preferred
Between two infeasible solutions, the one with more number of improved constraint violations is preferred.

If the solution remains feasible and unchanged for successive predefined number of iterations, and current feasible system objective is worse than the previous iteration feasible solution, accept the current solution. Similar to SA [4–7], this may help jump out of the local minima.

The detailed formulation of the above rule is explained below and further presented in window A of the constrained PC algorithm flowchart in Fig. 7.1.

If the current system objective $G(\mathbf{Y}^{[fav]})$ as well as the previous solution are infeasible, accept the current system objective $G(\mathbf{Y}^{[fav]})$ and corresponding $\mathbf{Y}^{[fav]}$ as current solution if the number of improved constraints is greater than or equal to μ, i.e. $C_{improved} \geq \mu$, and then the value of μ is updated to $C_{improved}$, i.e. $\mu = C_{improved}$.

If the current system objective $G(\mathbf{Y}^{[fav]})$ is feasible, and the previous solution is infeasible, accept the current system objective $G(\mathbf{Y}^{[fav]})$ and corresponding $\mathbf{Y}^{[fav]}$ as the current solution, and then the value of μ is updated to 0, i.e. $\mu = C_{improved} = 0$.

If the current system objective $G(\mathbf{Y}^{[fav]})$ is feasible and is not worse than the previous feasible solution, accept the current system objective $G(\mathbf{Y}^{[fav]})$ and corresponding $\mathbf{Y}^{[fav]}$ as the current solution.

If all the above conditions (a) to (c) are not met, then discard current system objective $G(\mathbf{Y}^{[fav]})$ and corresponding $\mathbf{Y}^{[fav]}$, and retain the previous iteration solution.

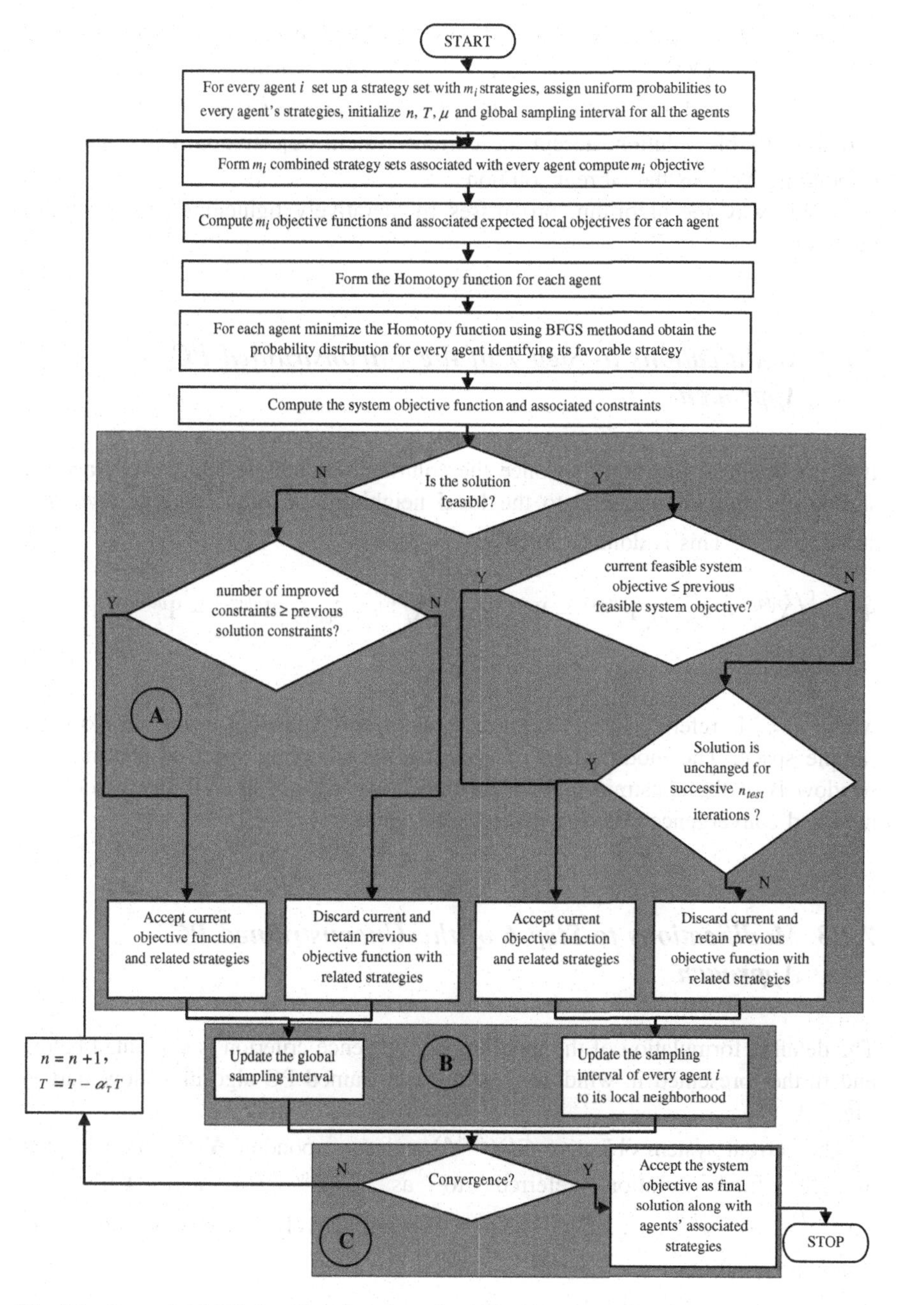

Fig. 7.1 Constrained PC algorithm flowchart (feasibility-based rule II)

If the solution remains feasible and unchanged for successive pre-specified n_{test} iterations i.e. $G(\mathbf{Y}^{[fav],n})$ and $G(\mathbf{Y}^{[fav],n-n_{test}})$ are feasible and $G(\mathbf{Y}^{[fav],n}) = G(\mathbf{Y}^{[fav],n-n_{test}})$, and the current feasible system objective is worse than the previous iteration feasible solution, accept the current system objective $G(\mathbf{Y}^{[fav]})$ and corresponding $\mathbf{Y}^{[fav]}$ as the current solution.

In this way the algorithm progresses by iteratively tightening the constraint violation.

7.1.2 Modifications to Step 7 of the Unconstrained PC Approach

It is important to mention that once the solution becomes feasible, every agent i shrinks its sampling interval to the local neighborhood of its current favorable strategy $X_i^{[fav]}$. This is done as follows:

$$\Psi_i \in \left[\left(X_i^{[fav]} - \lambda_{down}\left\|\Psi_i^{upper} - \Psi_i^{lower}\right\|\right), \left(X_i^{[fav]} + \lambda_{down}\left\|\Psi_i^{upper} - \Psi_i^{lower}\right\|\right)\right],$$
$$0 < \lambda_{down} \leq 1 \tag{7.4}$$

where λ_{down} is referred to as the interval factor corresponding to the shrinking of sample space. The modification of updating the sampling space is presented in window B of the constrained PC algorithm flowchart in Fig. 7.1. The associated modified convergence criterion is discussed below.

7.1.3 Modifications to Step 6 of the Unconstrained PC Approach

The detailed formulation of the modified convergence criterion is explained below and further presented in window C of the constrained PC algorithm flowchart in Fig. 7.1.

The current system objective $G(\mathbf{Y}^{[fav],n})$ and corresponding $\mathbf{Y}^{[fav],n}$ are accepted as the final solution referred to as $G(\mathbf{Y}^{[fav],final})$ and $\mathbf{Y}^{[fav],final} = \left\{X_1^{[fav],final}, X_2^{[fav],final}, \ldots, X_{N-1}^{[fav],final}, X_N^{[fav],final}\right\}$, respectively, if and only if any of the following conditions are satisfied.

If temperature $T = T_{final}$ or $T \to 0$

If maximum number of iterations exceeded

If there is no significant change in the system objective (i.e. $\left\| G(\mathbf{Y}^{[fav],n}) - G(\mathbf{Y}^{[fav],n-1}) \right\| \leq \varepsilon$) for successive considerable number of iterations.

7.2 The Sensor Network Coverage Problem (SNCP)

The sensor network plays a significant role in various strategic applications such as hostile and hazardous environmental and habitat exploration and surveillance, critical infrastructure monitoring and protection, situational awareness of battlefield and target detection, natural disaster relief, industrial sensing and diagnosis, biomedical health monitoring, seismic sensing, etc. [8–16]. The deployment or positioning of the individual sensor directly affects the coverage, detection capability, connectivity and associated communication cost and resource management of the entire network [10, 11, 17, 18]. According to [12, 17, 19–21], coverage is the important performance metric that quantifies the quality and effectiveness of the surveillance/monitoring provided by the sensor network. This highlighted the requirement for an effective sensor deployment algorithm to optimize the coverage of the entire network [12, 17, 18]. Furthermore, according to [22, 23] the connected coverage is important by which all the sensors in the network can communicate with one another over multiple hops.

The sensor network coverage can be classified as deterministic or stochastic coverage [24]. The deterministic coverage refers to the static deployment of the sensors over a predefined Field of Interest (FoI) [25–29]. It includes the uniform coverage of the FoI as well as weighted coverage of certain section of the FoI [20, 30]. The stochastic coverage refers to the random deployment of the sensors over the FoI. The sensor positions are selected based on the uniform, Gaussian, Poisson or any other problem specific distribution [20]. The deterministic coverage may provide worst-case performance which may not be achievable by stochastic coverage [12, 21]. According to [31], the coverage can also be classified into three types. The first is blanket coverage, which represents the static arrangement of the sensors to maximize the total detection area; the second is barrier coverage, which aims to arrange the sensors to form a barrier in order to minimize the probability of undetected penetration/intrusion through the barrier [21, 31]; and the third is sweep coverage which is a moving barrier of sensors that aims to maximize the number of detections per unit time and minimize the number of missed detections per unit area. The blanket coverage can be further classified into two types [8]. The first is point-set coverage, which aims to cover a set of discrete points scattered over a certain field and the second is FoI coverage, which aims to cover a certain region to the maximum possible extent.

This following section addresses the problem of deployment of a set of homogeneous (identical characteristics) sensors over a certain predefined FoI in order to achieve the maximum possible deterministic connected blanket coverage using the method of PC.

7.2.1 Formulation of the SNCP

The SNCP addressed here is as follows: given a set of z homogeneous sensors with equal and fixed sensing range r_s for every sensor $i \in \{1, 2, \ldots, z\}$, and the sensing area of every sensor i modeled as a circle with the sensor located at its centre; find the deployment of all the sensors to collectively cover the maximum possible area (hereafter referred to as collective coverage $A_{collective}$) of a square FoI without overlapping one another and without exceeding the boundaries of the square FoI, and importantly maintaining connectivity between them. The connectivity here refers to the constraint which limits the distance $d(i,j)$ between two adjacent sensors i and j to a maximum predefined value γ, i.e. $d(i,j) \leq \gamma$, $i \sim j$, $i \neq j$, $i,j = 1, 2, \ldots, z$, [22, 23] where adjacency is represented by the symbol '$\sim$'. The network is considered connected if there is at least one connected path between each and every pair of sensors. As mentioned previously, it ensures that all the sensors in the network can communicate with one another over multiple hops. Variations to this connectivity constraint based on the number of sensors required [22], and the specific deployment pattern under consideration [22, 23], can be found in the literature. An example of a SNCP with three sensors ($z = 3$) is illustrated in Fig. 7.2 in which the distance $d(1, 2)$ between adjacent sensors 1 and 2 and $d(2, 3)$ between adjacent sensors 2 and 3 is represented by dashed lines with arrows at both the ends.

In order to achieve the above mentioned objective of maximum possible collective coverage $A_{collective}$ of the square FoI and yet maintaining connectivity of the sensors the system objective is the area $A_{\square}$ of an enclosing rectangle for which the sensing areas of all the sensors can be fitted within. The sensors are considered as

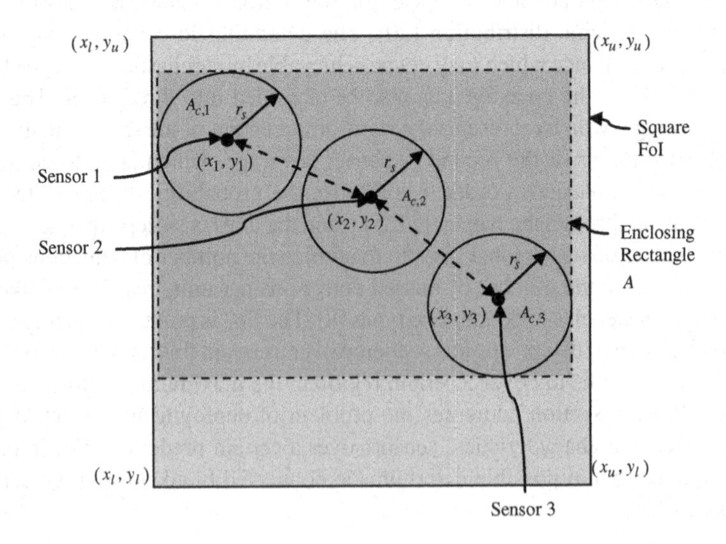

Fig. 7.2 An example of a sensor network coverage problem

autonomous agents and are assigned the individual x coordinates and y coordinates as their strategy sets. The SNCP is formulated as follows:

Minimize

$$A_{\square} = ((\max(x_1, x_2, \ldots, x_i, \ldots, x_z) + r_s) - (\min(x_1, x_2, \ldots, x_i, \ldots, x_z) - r_s)) \times ((\max(y_1, y_2, \ldots, y_i, \ldots, y_z) + r_s) - (\min(y_1, y_2, \ldots, y_i, \ldots, y_z) - r_s)) \quad (7.5)$$

Subject to

$$d(i,j) \geq 2r_s, \quad i,j = 1, 2, \ldots, z, \quad i \neq j \quad (7.6)$$

$$x_i - r_s \geq x_l \quad (7.7)$$

$$x_i + r_s \leq x_u \quad (7.8)$$

$$y_i - r_s \geq y_l \quad (7.9)$$

$$y_i + r_s \leq y_u \quad (7.10)$$

Table 7.1 Summary of results

Sr No	Particulars	Variation 1	Variation 2		
1	Cases	–	Case 1	Case 2	Case 3
2	Number of sensors (z)	5	5	10	20
3	The sensing range (r_s)	0.5	1.2	1	0.6
4	Average collective coverage	3.927	18.5237	19.4856	16.3631
5	Minimum and maximum collective coverage	3.9270, 3.9270	18.0920, 18.7552	17.5427, 20.8797	15.5347, 17.3377
6	Standard deviation associated with collective coverage	0.0000	0.1687	1.1837	1.2217
7	Average area of the enclosing rectangle	5.8311	34.3014	49.0938	39.3480
8	Minimum and maximum area of the enclosing rectangle	5.7046, 5.9750	33.0448, 39.7099	44.7135, 52.6277	34.1334, 43.8683
9	Standard deviation associated with the area of enclosing rectangle	0.1040	1.9899	2.6995	2.8829
10	Average CPU time (approx.)	20 mins	1 h	2 h	3.5 h
11	Average number of function evaluations	90,417	315,063	1,172,759	3,555,493

$$i = 1, 2, \ldots, z$$

$$d(i,j) \leq \gamma \qquad i \sim j, \quad i \neq j, \quad i,j = 1, 2, \ldots, z \tag{7.11}$$

where

$A_{\square}$ = area of the enclosing rectangle
z = number of sensors
r_s = sensing range of every sensor $i \in \{1, 2, \ldots, z\}$
$x_i,\ y_i$ = x and y coordinates of sensor i (or x and y coordinates of the center of the circle i)
$x_l,\ y_l$ = x and y coordinates of the lower left corner of the square FoI
$x_u,\ y_u$ = x and y coordinates of the upper right corner of the square FoI
$d(i,j) = \sqrt{(x_i - x_j)^2 + (y_i - y_j)^2}$ = distance between two sensors i and j

The constraints in Eq. (7.6) are to prevent overlapping of the sensor circles and the constraints in Eqs. (7.7–7.10) are to prevent the sensor circles from exceeding the FoI boundaries. The constraints in Eq. (7.11) are the connectivity constraints and are checked in every iteration using the well known Depth First Search (DFS) algorithm [32] only when all the other constraints are satisfied.

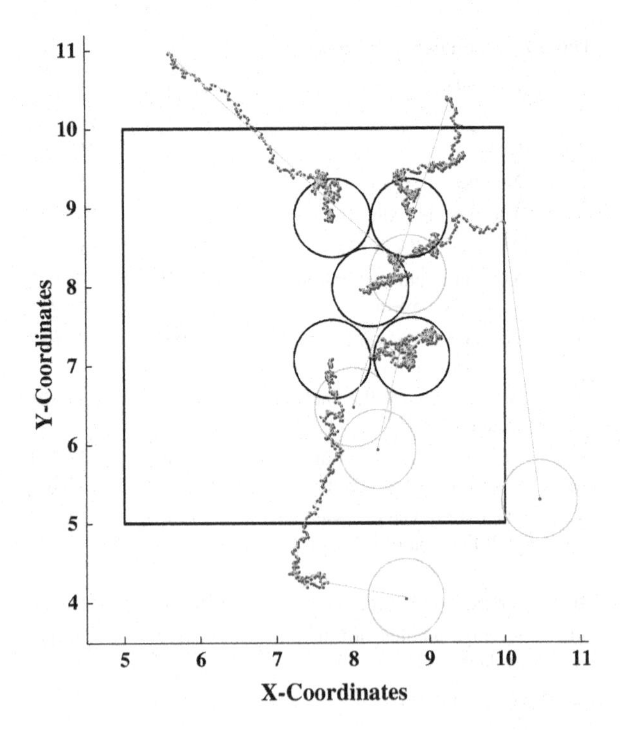

Fig. 7.3 Solution history for variation 1

The associated collective coverage $A_{collective}$ is calculated as follows:

$$A_{collective} = \sum_{i=1}^{z} A_{c,i} \tag{7.12}$$

where $A_{c,i}$ = FoI coverage achieved by sensor i.

In the three sensor example illustrated in Fig. 7.2, the individual sensor coverage area is $A_{c,i} = \pi r_s^2$ and the collective coverage is $A_{collective} = \sum_{i=1}^{3} A_{c,i} = 3\pi r_s^2$. However, there can be many possible instances where the sensing areas of the

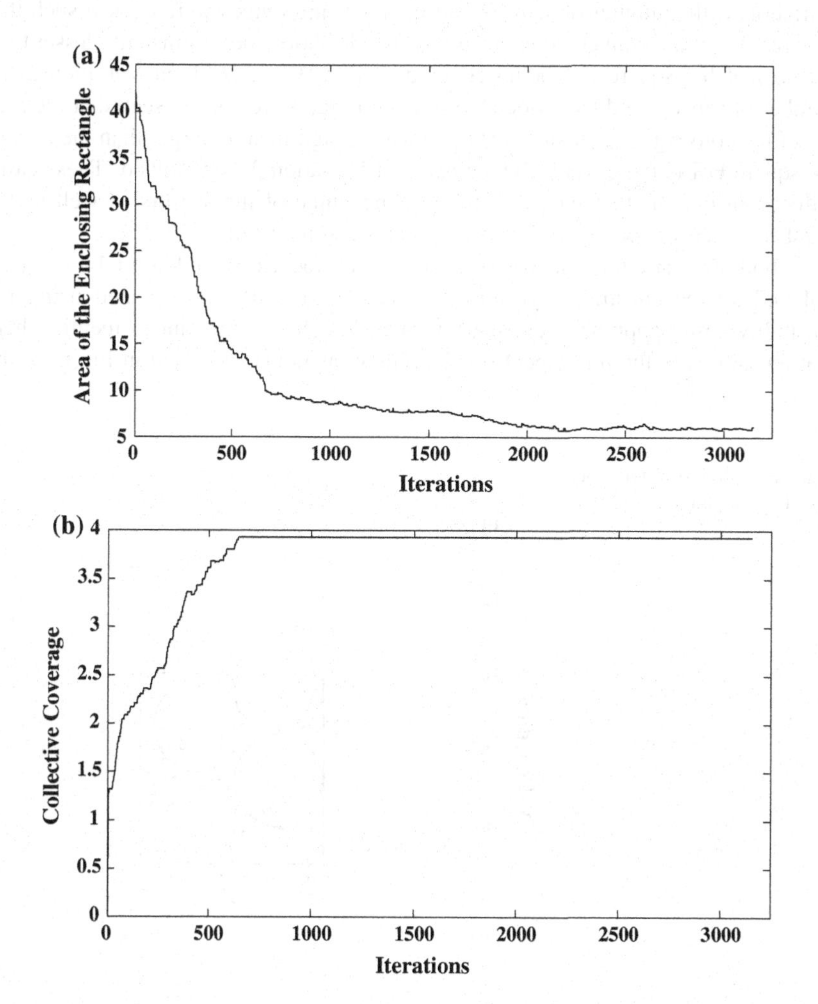

Fig. 7.4 Convergence of the area of the enclosing rectangle and collective coverage for variation 1. **a** Convergence of the area of the enclosing rectangle. **b** Convergence of the collective coverage

sensors may not be completely within the FoI, and the required computation for these different individual sensor coverage cases are discussed in Appendix D.

7.2.2 Variations of the SNCP Solved

In solving the SNCP using constrained PC approach, the sensors were considered as autonomous agents. These sensors were assigned the individual x coordinates and y coordinates as their strategy sets. Two variations of the SNCP were solved. In Variation 1, the number of sensors and their sensing ranges were chosen such that the area $A_{\square}$ of the converged enclosing rectangle should be significantly lesser than the area of the square FoI being covered. Unlike Variation 1, in Variation 2 the number of sensors and the associated sensing range were chosen such that the area $A_{\square}$ of the converged enclosing rectangle will be significantly larger than the area of the square FoI being covered. Three cases of Variation 2 were solved. These cases differ from one another based on the sensing range of the sensors as well as the number of sensors being deployed to cover the square FoI.

In both the variations, sensors were randomly located in-and-around the square FoI and all the constraints presented in Eqs. (7.6–7.10) were treated using the feasibility-based approach described in Sect. 7.1. It is important to mention here that on satisfying the inter-agent overlapping constraints presented in Eq. (7.6) the

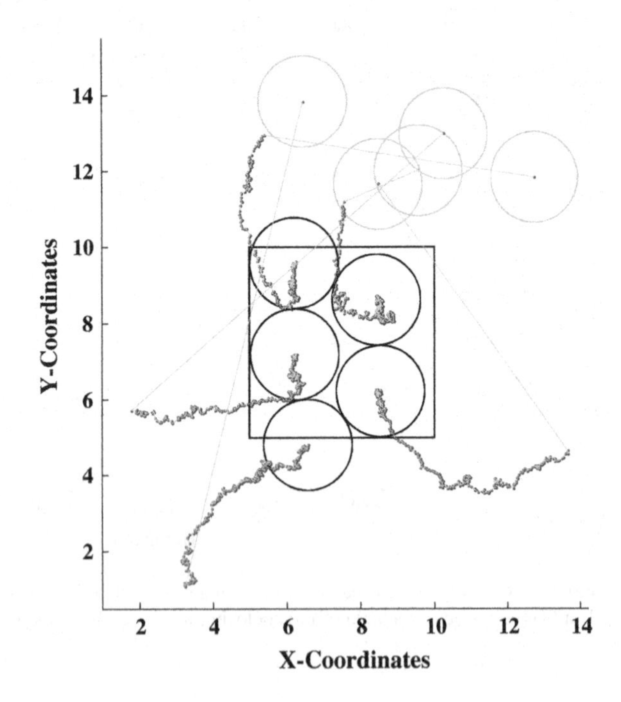

Fig. 7.5 Solution history for case 1 of variation 2

sampling space of every individual agent was shrunk to its local neighborhood. The approach of shrinking the sampling space is discussed in Eq. (7.4).

The constrained PC algorithm solving both the variations was coded in MATLAB 7.8.0 (R2009A) and the simulations were run on a Windows platform using an Intel Core 2 Duo, 3 GHz processor speed and 3.25 GB memory capacity. Furthermore, for both the variations the set of parameters chosen was as follows:

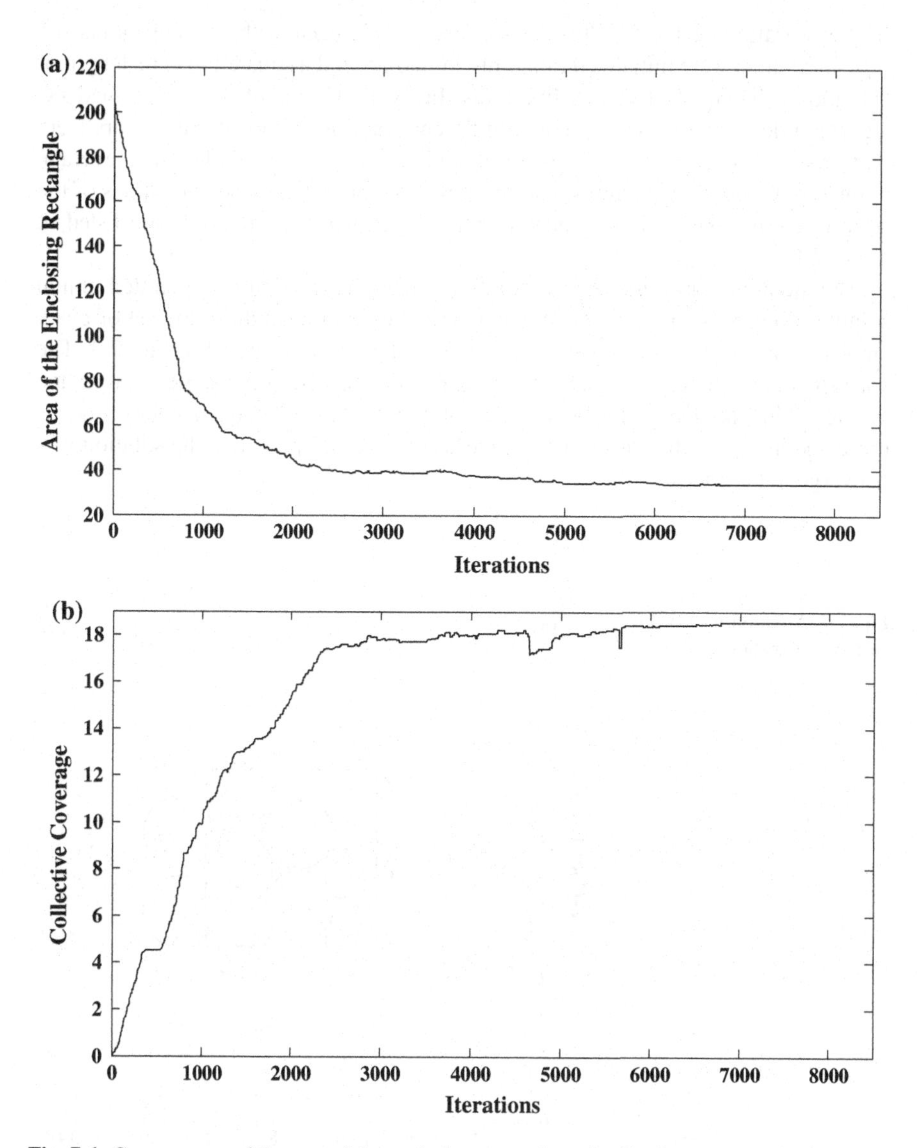

Fig. 7.6 Convergence of the area of the enclosing rectangle and collective coverage for case 1 of variation 2. **a** Convergence of area of the enclosing rectangle. **b** Convergence of the collective coverage

(a) individual agent sample size $m_i = 5$, (b) the shrinking interval factor $\lambda_{down} = 0.05$, (c) number of test iterations $n_{test} = 20$. In addition, the maximum allowable distance between two adjacent sensors γ was chosen as $0.2 + 2r_s$ units.

7.2.2.1 Variation 1

In this variation of the SNCP, five sensors ($z = 5$) each with sensing radius 0.5 ($r_s = 0.5$) units were initialized randomly in-and-around a square FoI. The length of the side of the square FoI was five units. In total, 10 runs of the constrained PC algorithm described in Sect. 7.1 with different initial locations of the sensors were performed. The summary of results including the average CPU time, average number of function evaluations, the minimum, maximum and average values of the collective coverage achieved and the area of the enclosing rectangle are listed in Table 7.1.

The iterative progress of the solution including randomly generated initial solution (the set of gray circles) and converged optimum solution (the set of black circles) from one of the instances solving Variation 1 is presented in Fig. 7.3. The corresponding convergence plot of the area of the enclosing rectangle is presented in Fig. 7.4a and the convergence plot of the associated collective coverage is presented in Fig. 7.4b. It is clear from these convergence plots that the solution was converged at iteration 2,198.

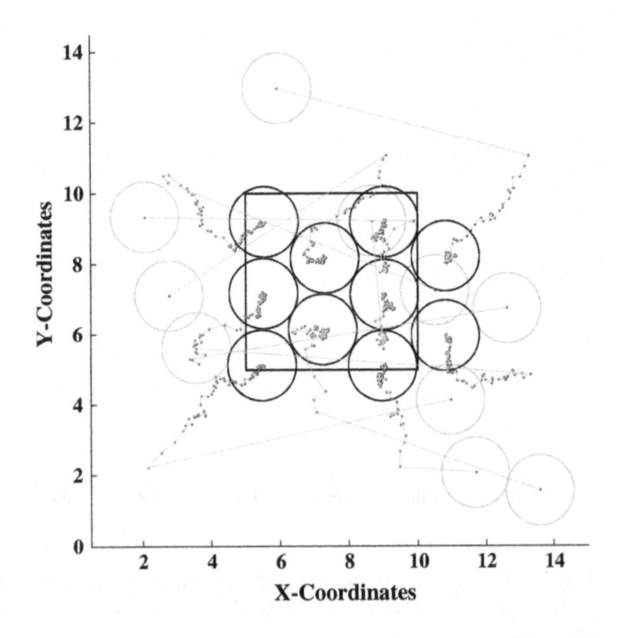

Fig. 7.7 Solution history for case 2 of variation 2

7.2.2.2 Variation 2

Three distinct cases of Variation 2 were solved. Case 1 includes five sensors ($z = 5$) each with sensing radius 1.2 ($r_s = 1.2$), Case 2 includes ten sensors ($z = 10$) each with sensing radius 1 ($r_s = 1$) and Case 3 includes twenty sensors ($z = 20$) each with sensing radius 0.6 ($r_s = 0.6$) units. In each of the three cases, sensors were initialized randomly in-and-around the square FoI. The length of the side of the square FoI was five units. In total, 10 runs of the constrained PC algorithm

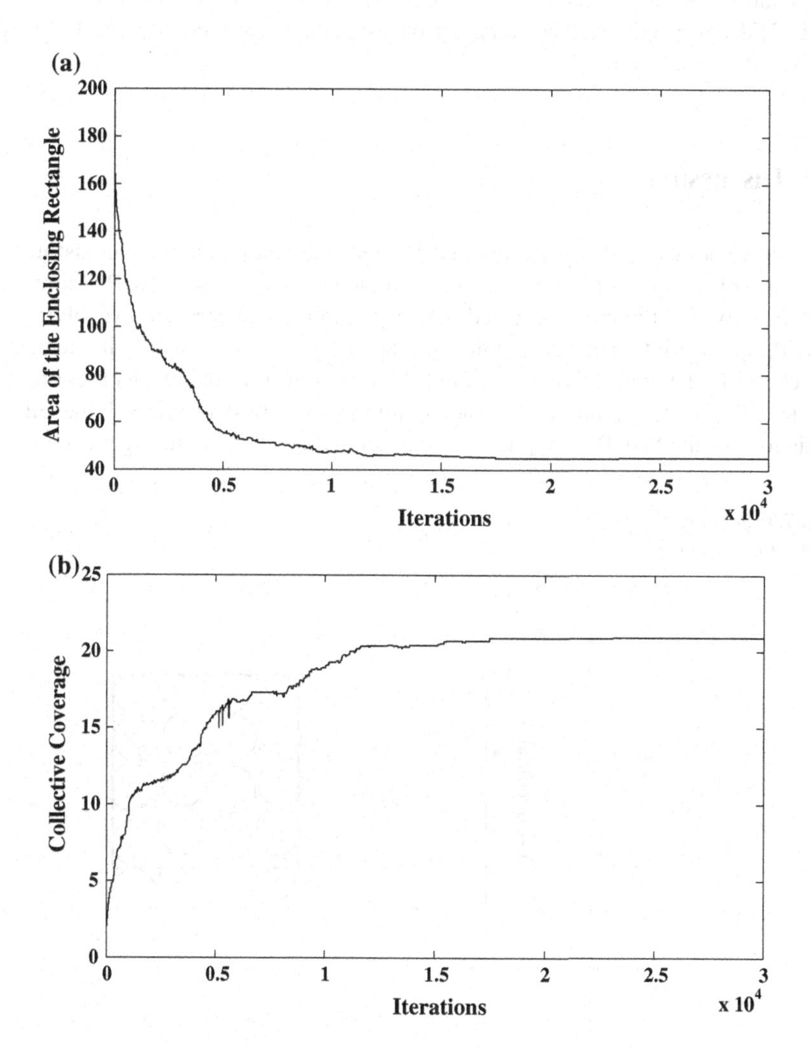

Fig. 7.8 Convergence of the area of the enclosing rectangle and collective coverage for case 2 of variation 2. **a** Convergence of the area of the enclosing rectangle. **b** Convergence of the collective coverage

described in Sect. 7.1 with different initial locations of the sensors were conducted for each of the three cases.

The iterative progress of the solution including randomly generated initial solution (the set of gray circles) and the converged optimum solution (the set of black circles) from one of the instances each of Case 1, 2 and 3 are illustrated in Figs. 7.5, 7.7 and 7.9, respectively. In addition, the corresponding convergence plots of the area of the enclosing rectangle are presented in Figs. 7.6a, 7.8a and 7.10a, and the convergence plots of the collective coverage are presented in Figs. 7.6b, 7.8b and 7.10b, respectively. It is clear from these convergence plots that the solution for Case 1, Case 2 and Case 3 was converged at iteration 6,725, 17,484 and 42,218, respectively. The summary of results corresponding to Case 1, 2 and 3 are listed in Table 7.1.

7.3 Discussion

The above solutions using constrained PC indicated that it could successfully be used to solve constrained optimization problems such as the SNCP. Moreover, connectivity constraint was satisfied in every run of the constrained PC solving the SNCP. It is worth to mention that the DFS algorithm was successfully implemented for checking the connectivity constraint. It is evident from the results presented in Table 7.1, that the approach produced sufficiently robust results solving all the variations of the SNCP. It implies that the rational behavior of the agents could be

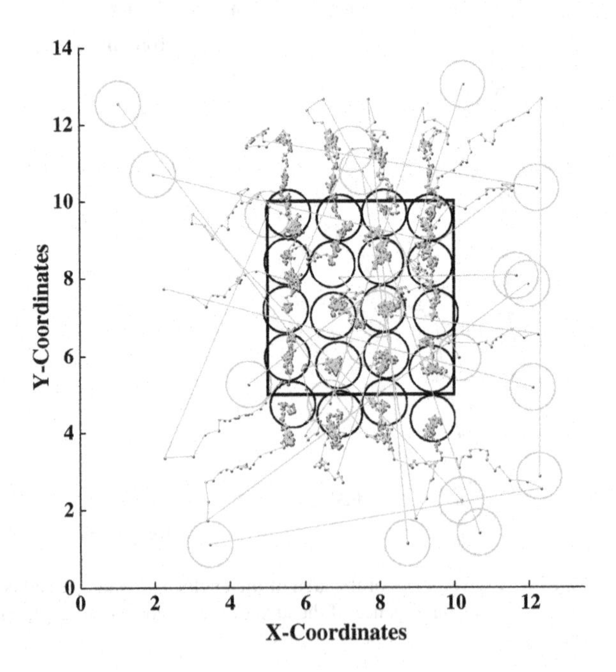

Fig. 7.9 Solution history for case 3 of variation 2

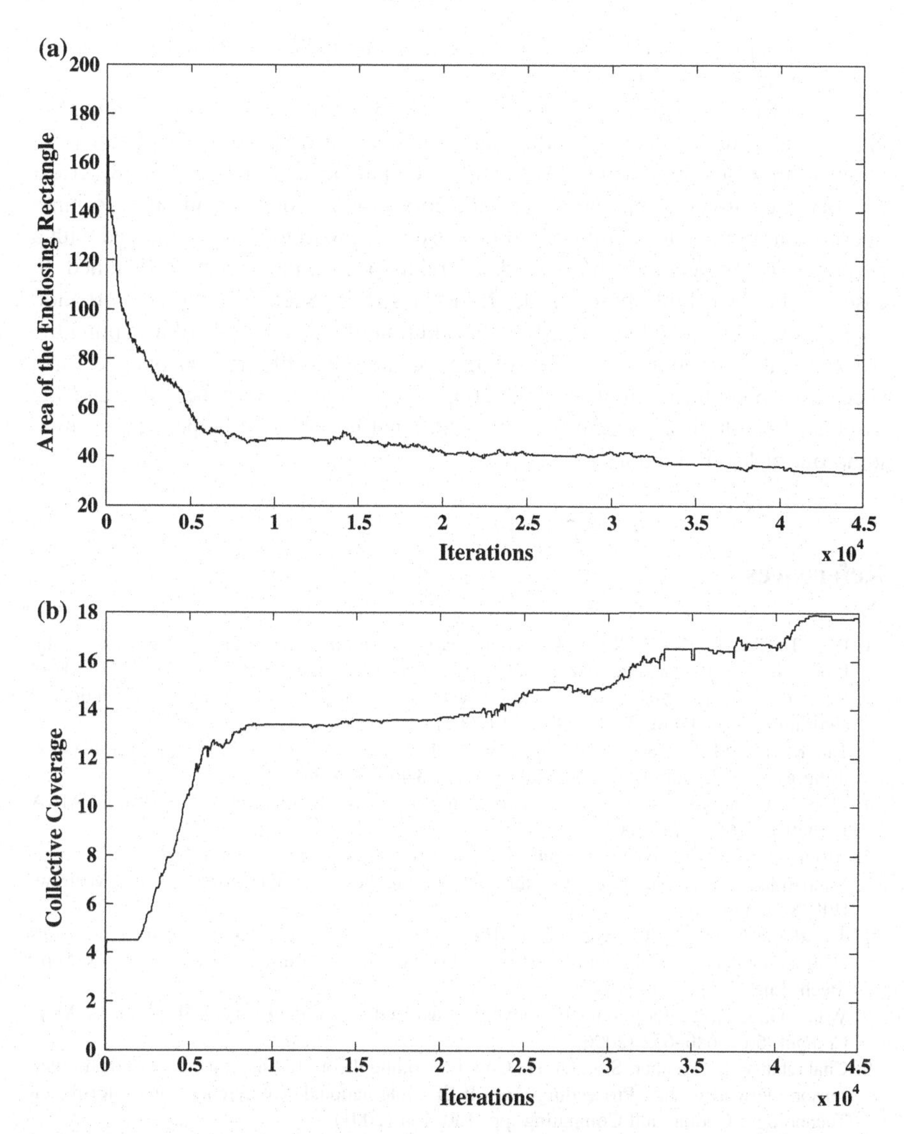

Fig. 7.10 Convergence of the area of the enclosing rectangle and collective coverage for case 3 of variation 2. **a** Convergence of the area of the enclosing rectangle. **b** Convergence of the collective coverage

successfully formulated and demonstrated. It is important to highlight that the distributed nature of the PC approach will allow the total number of function evaluations to be equally divided among the agents of the system. This can be made practically evident by implementing the PC approach on a real distributed platform assigning separate workstations carrying out the computations independently. These advantages along with the directly incorporated uncertainty using the real valued probabilities treated as variables suggest that PC can potentially be applied to real world complex problems.

In addition, the feasibility-based rule in [33–39] suffered from maintaining the diversity and further required additional techniques such as niching [33], SA [34], modified mutation approach [37, 38], and several associated trials in [35–38], etc. It may require further computations and memory usage. Moreover, in order to jump out of local minima, feasibility-based Rule I was required to be accompanied with a perturbation approach which required several associated parameters to be tuned. In case of the Feasibility-based Rule II discussed in Sect. 7.1, the perturbation approach was completely replaced by the addition of a generic rule of accepting the worse feasible solution when the solution remains feasible and unchanged for a successive predefined number of iterations. From the convergence plot and the associated results it is evident that this additional generic rule helped the solution jump out of local minima.

References

1. Ray, T., Tai, K., Seow, K.C.: An evolutionary algorithm for constrained optimization. In: Proceedings of the Genetic and Evolutionary Computation Conference, pp. 771–777 (2000)
2. Ray, T., Tai, K., Seow, K.C.: Multiobjective design optimization by an evolutionary algorithm. Eng. Optim. **33**(4), 399–424 (2001)
3. Tai, K., Prasad, J.: Target-matching test problem for multiobjective topology optimization using genetic algorithms. Struct. Multi. Optim. **34**(4), 333–345 (2007)
4. Moh, J., Chiang, D.: Improved simulated annealing search for structural optimization. AIAA J. **38**(10), 1965–1973 (2000)
5. Arora, J.S.: Introduction to Optimum Design. Elsevier Academic Press, (2004)
6. Vanderplaat, G.N.: Numerical Optimization Techniques for Engineering Design. Mcgraw-Hill, New York (1984)
7. Theodoracatos, V.E., Grimsley, J.L.: The optimal packing of arbitrarily-shaped polygons using simulated annealing and polynomial-time cooling schedules. Comput. Methods Appl. Mech. Eng. **125**, 53–70 (1995)
8. Wang, G., Cao, G.: La porta T.F.: movement-assisted sensor deployment. IEEE Trans. Mob. Comput. **5**(6), 640–652 (2006)
9. Chakrabarty, K., Iyengar, S.S., Qi, H., Cho, E.: Coding theory for target location in distributed sensor networks. In: Proceedings of IEEE International Conference on Information Technology: Coding and Computing, pp. 130–134 (2001)
10. Chakrabarty, K., Iyengar, S.S., Qi, H., Cho, E.: Grid coverage for surveillance and target location in distributed sensor networks. IEEE Trans. Comput. **51**(12), 1448–1453 (2002)
11. Yick, J., Mukherjee, B., Ghosal, D.: Wireless sensor network survey. Comput. Netw. **52**(12), 2292–2330 (2008)
12. Castillo-Effen, M., Quintela, D.H., Jordan, R., Westhoff, W., Moreno, W.: Wireless sensor networks for flash-flood alerting. In: Proceedings of 5th IEEE International Caracas Conference on Devices, Circuits, and Systems, Dominican Republic, pp. 142–146 (2004)
13. Gao, T., Greenspan, D., Welsh, M., Juang, R.R., Alm, A.: Vital signs monitoring and patient tracking over a wireless network. In: Proceedings of 27th IEEE EMBS Annual International Conference, pp.102–105 (2005)

14. Lorincz, K., Malan, D., Fulford-Jones, T.R.F., Nawoj, A., Clavel, A., Shnayder, V., Mainland, G., Welsh, M., Moulton, S.: Sensor networks for emergency response: challenges and opportunities. IEEE Pervasive Comput. **3**(4), 16–23 (2004)
15. Wener-Allen, G., Lorincz, K., Ruiz, M., Marcillo, O., Johnson, J., Lees, J., Walsh, M.: Deploying a wireless sensor network on an active volcano. IEEE Internet Comput. **10**(2), 18–25 (2006)
16. Zou Y, Chakrabarty, K.: Sensor Deployment and Target Localization Based on Virtual Forces. In: Proceedings of the IEEE INFOCOM 2003, pp. 1293–1303 (2003)
17. Lee, J., Jayasuriya, S.: Deployment of mobile sensor networks with discontinuous dynamics. In: Proceedings of International Federantion of Automatic Control, pp. 10409–10414 (2008)
18. Lazos, L., Poovendran, R.: Stochastic coverage in heterogeneous sensor networks. ACM Trans. Sens. Netw. **2**(3), 325–358 (2006)
19. Meguerdichian, S. Koushanfar, F., Potkonjak, M., Srivastava, M.B.: Coverage problems in wireless ad-hoc sensor networks, In: Proceedings of IEEE International Conference on Computer Communications, pp. 1380–1387 (2001)
20. Lazos, L., Poovendran, R.: Detection o mobile coverage targets on the plane and in space using heterogeneous sensor networks. Wireless Netw. **15**(5), 667–690 (2009)
21. Lazos, L., Poovendran, R.: Coverage in heterogeneous sensor networks. In: Proceedings of International Symposium of Modeling and Optimization in Mobile, Ad-Hoc and Wireless Networks, pp. 1–10, (2006)
22. Xing, G., Wang, X., Zhang, Y., Lu, C., Pless, R., Gill, C.: Integrated coverage and connectivity configuration in wireless sensor networks. ACM Trans. Sens. Netw. **1**(1), 36–72 (2005)
23. Iyengar, R., Kar, K., Banerjee, S.: Low-coordination topologies for redundancy in sensor networks. In Proceedings of 6th ACM Annual International Symposium on Mobile Ad-Hoc Networking and Computing, pp. 332–342 (2005)
24. Clouqueur, T., Phipatanasuphorn, V., Ramanathan P., Saluja, K.K.: Sensor deployment strategy for target detection. In: Proceedings of 1st ACM International Workshop on Wireless Sensor Networks and Applications, pp. 42–48 (2002)
25. Zhao, F., Guibas, L.: Wireless Sensor Networks. Morgan Kaufmann, Burlington (2004)
26. Fortune, S., Du, D.: Hwang, F.: Voronoi Diagrams and Delaunay Triangulations. CRC Press, Handbook of Euclidean Geometry and Computers (1997)
27. Liu, B., Towsley, D.: A study of the coverage of large-scale sensor networks. In: Proceedings of IEEE International Conference on Mobile Ad-Hoc and Sensor Systems, pp. 475–483 (2004)
28. Poduri, S., Sukhatme, G.S.: Constrained coverage for mobile sensor networks, In: Proceedings of IEEE International Conference on Robotics and Automation, pp. 165–172 (2004)
29. Marengoni, M., Draper, B.A., Hanson, A., Sitaraman, R.A.: System to place observers on a polyhedral terrain in polynomial time. Image Vis. Comput. **18**(10), 773–780 (1996)
30. Gage, D.W.: Command control for many-robot systems. Inmanned Syst. Mag. **10**(4), 28–34 (1992)
31. Bai, X., Kumar, S., Xuan, D., Yun, Z., Lai, T.H.: Deploying wireless sensors to achieve both coverage and connectivity. In: Proceedings of International Symposium on Mobile Ad Hoc Networking and Computing, pp. 131–142 (2006)
32. Lugar, G.F.: Artificial Intelligence: Structures and Strategies for Complex Problem Solving. Addison-Wesley, Boston (2005)
33. Deb, K.: An efficient constraint handling method for genetic algorithms. Comput. Methods Appl. Mech. Eng. **186**, 311–338 (2000)
34. He, Q., Wang, L.: A hybrid particle swarm optimization with a feasibility-based rule for constrained optimization. Appl. Math. Comput. **186**, 1407–1422 (2007)
35. Sakthivel, V.P., Bhuvaneswari, R., Subramanian, S.: Design optimization of three-phase energy efficient induction motor using adaptive bacterial foraging algorithm. Comput. Math. Electr. Electron. Eng. **29**(3), 699–726 (2010)
36. Kaveh, A., Talatahari, S.: An amproved ant colony optimization for constrained engineering design problems. Comput. Aided Eng. Softw. **27**(1), 155–182 (2010)

37. Bansal, S., Mani, A., Patvardhan, C.: Is stochastic ranking really better than feasibility rules for constraint handling in evolutionary algorithms? In: Proceedings of World Congress on Nature and Biologically Inspired Computing, pp. 1564–1567 (2009)
38. Gao, J., Li, H., Jiao, Y.C.: Modified differential evolution for the integer programming problems. In: Proceedings of International Conference on Artificial Intelligence, pp. 213–219 (2009)
39. Ping, W., Xuemin, T.: A hybrid DE-SQP algorithm with switching procedure for dynamic optimization. In: Proceedings of Joint 48th IEEE Conference on Decision and Control and 28th Chinese Control Conference, pp. 2254–2259 (2009)

Appendix A
Analogy of Homotopy Function to Helmholtz Free Energy and Deterministic Annealing

1. The Homotopy function $J_i(q(\mathbf{X}_i), T)$ in Eq. (2.10) is analogous to the Helmholtz free energy in statistical physics. The Helmholtz free energy equation can be represented as follows:

$$L = D - TS \tag{A.1}$$

where L is Helmholtz free energy available to do the work or to create a system when the environment is at temperature T. The term D is referred to as internal energy and S is the entropy of the physical system. The term TS is the spontaneous energy that can be transferred from the environment to the system.

As Helmholtz free energy L is actually the measure of the amount of internal energy D needed to create the system, achieving thermal equilibrium is nothing but minimizing the free energy L at temperature T, and hence directly minimizing the internal energy D. The procedure involves a series of iterations gradually reducing the entropy level of the system. This suggests the framework of an annealing schedule, i.e. start at a high temperature $T \gg 0$ or $T = T_{initial}$ and successively reducing the temperature and the entropy level, achieving the thermal equilibrium at every successive temperature drop until $T \to 0$ or $T = T_{final}$. This process avoids the local minima of internal energy D and reaches the global minima at $T \to 0$ or $T = T_{final}$.

2. The analogy of the Homotopy function $J_i(q(\mathbf{X}_i), T)$ to Helmholtz free energy L in statistical physics is the motivation behind minimizing the collection of system objectives $\sum_{r=1}^{m_i} G(\mathbf{Y}_i^{[r]})$. As thermal equilibrium needs to be achieved to minimize the free energy L of the physical system given in Eq. (A.1), similarly, the equilibrium needs to be achieved to minimize the Homotopy function $J_i(q(\mathbf{X}_i), T)$ and that is referred to as Nash equilibrium.
3. The careful annealing of physical systems ensures the convergence to the global minimum of the energy D. One of the most suitable approaches to deal with the Homotopy function such as $J_i(q(\mathbf{X}_i), T)$ and eventually the collection of system

© Springer International Publishing Switzerland 2015

A.J. Kulkarni et al., *Probability Collectives*, Intelligent Systems Reference Library 86, DOI 10.1007/978-3-319-16000-9

objectives $\sum_{r=1}^{m_i} G(\mathbf{Y}_i^{[r]})$ having many possible local minima is referred to as Deterministic Annealing (DA).

4. More specifically, the DA refers to the procedure discussed in A.1 which suggests the minimization of the Homotopy function $J_i(q(\mathbf{X}_i), T) = \sum_{r=1}^{m_i} E\left(G(\mathbf{Y}_i^{[r]})\right) - TS_i$, $i = 1, \ldots N$ in a series of iterations gradually reducing the effect of entropy function S_i, i.e. minimize the Homotopy function $J_i(q(\mathbf{X}_i), T)$ at a high temperature $T \gg 0$ or $T = T_{initial}$ at which the convex entropy function S_i dominates the multimodal function $\sum_{r=1}^{m_i} G(\mathbf{Y}_i^{[r]})$; successively/iteratively reduce the temperature and the effect of entropy function S_i and minimize the Homotopy function $J_i(q(\mathbf{X}_i), T)$ in every corresponding iteration. This slowly reduces the effect of entropy function S_i and gradually makes the collection of system objectives $\sum_{r=1}^{m_i} G(\mathbf{Y}_i^{[r]})$ to dominate. This helps achieving the Nash equilibrium at every successive temperature drop until $T \to 0$ or $T = T_{final}$ as well as avoids the possible local minima of the collection of system objectives $\sum_{r=1}^{m_i} G(\mathbf{Y}_i^{[r]})$ and helps to reach in a very global minimum of it at $T \to 0$ or $T = T_{final}$. This is important to mention that at $T \to 0$ or $T = T_{final}$ the effect of entropy function S_i will be negligible.

Appendix B
Nearest Newton Descent Scheme

In this scheme, the quadratic approximation of the Hessian of Homotopy function $J_i(q(\mathbf{X}_i), T)$ in Eq. (2.10) is carried out by every agent i using the probability space formed from the coupling of the individual agent's probability spaces rather than using the individual agent's probability distribution $q(\mathbf{X}_i)$. It is worth to mention that the Hessian in this scheme is positive definite. The important benefit is that it converts the Homotopy function $J_i(q(\mathbf{X}_i), T)$ into a convex function. The gradient and Hessian of the Homotopy function in Eq. (2.10) is used to find the individual agent probability distribution.

The simplified resulting probability variable update rule (descent rule) for each strategy r of agent i, referred to as the 'k-update' rule which minimizes the Homotopy function in Eq. (2.10), is represented below:

$$q(X_i^{[r]})^{k+1} \leftarrow q(X_i^{[r]})^k - \alpha_{step}\, q(X_i^{[r]})^k k_{r.update} \tag{B.1}$$

where

$$k_{r.update} = \frac{(Contribution\ of\ X_i^{[r]})^k}{T} + S_i^k + \log_2(q(X_i^{[r]})^k)$$

and

$$\left(Contribution\ of\ X_i^{[r]}\right)^k = \left(E\left(G(\mathbf{Y}_i^{[r]})\right)\right)^k - \left(\sum_{r=1}^{m_i} E\left(G(\mathbf{Y}_i^{[r]})\right)\right)^k$$

where k is the corresponding update number and S_i^k is the corresponding entropy. The updating procedure is applied as follows:

(a) Set iteration number $k = 1$, maximum number of updates k_{final}, and step size α_{step} $(0 < \alpha_{step} \leq 1)$. The values of k_{final} and α_{step} are held constant throughout the optimization and chosen based on preliminary trials of the algorithm.

(b) Update every agent's probability distribution $q(\mathbf{X}_i)^{(k)}$ using the update rule in Eq. (B.1).

© Springer International Publishing Switzerland 2015
A.J. Kulkarni et al., *Probability Collectives*, Intelligent Systems Reference Library 86, DOI 10.1007/978-3-319-16000-9

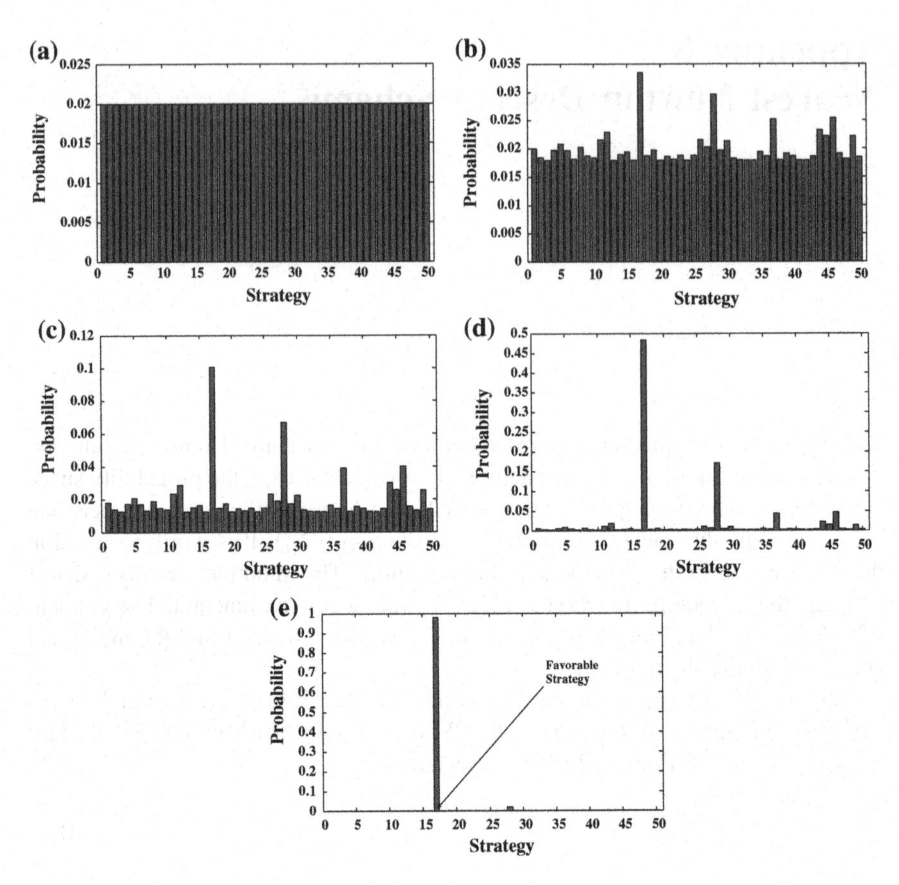

Fig. B.1 Solution history of probability distribution using nearest newton descent scheme

(c) If $k \geq k_{final}$, stop, else update $k = k + 1$ and return to (b).

For each agent i, the above optimization process converges to a probability variable vector $q(\mathbf{X}_i)$ which can be seen as the individual agent's probability distribution clearly distinguishing every strategy's contribution towards the minimization of the expected collection of system objectives $\sum_{r=1}^{m_i} E\left(G(\mathbf{Y}_i^{[r]})\right)$. In other words, for every agent i, if strategy r contributes the most towards the minimization of the objective than other strategies, its corresponding probability certainly increases by some amount more than those for the other strategies' probability values, and so strategy r is distinguished from the other strategies. Such a strategy is referred to as a favorable strategy $X_i^{[fav]}$.

The initial uniform probability distribution, the intermediate iteration distributions, and the converged probability distribution clearly distinguishing the contribution of every strategy from one of the runs solving the tension/compression spring design problem discussed in Chap. 4 are illustrated in Fig. B.1.

Appendix C
Broyden-Fletcher-Goldfarb-Shanno (BFGS) Method for Minimizing the Homotopy Function

The minimization of the Homotopy function given in Eq. (2.10) was carried out using a suitable second order optimization technique such as Broyden-Fletcher-Goldfarb-Shanno (BFGS) method. The approximated Hessian in this method is positive definite. Moreover, the updating of the Hessian also preserves the positive definiteness. The BFGS method minimizing the Homotopy function in Eq. (2.10) is discussed below.

1. Set BFGS iteration counter $k = 1$, BFGS maximum number of iterations v, and step size α_{step} $(0 < \alpha_{step} \leq 1)$. The value of α_{step} is held constant throughout the optimization and chosen based on the preliminary trials of the algorithm.

1.1 Initialize the convergence criterion $q(\mathbf{X}_i)^k - q(\mathbf{X}_i)^{k-1} \leq \varepsilon_2$. The convergence parameter $\varepsilon_2 = 0.0001$ is equal for all the N agents.

1.2 Initialize the Hessian $\mathbf{H}_i^k$ to a positive definite matrix, preferably identity matrix $\mathbf{I}$ of size $m_i \times m_i$.

1.3 Initialize the probability variables as follows:

$$q(\mathbf{X}_i)^k = \left\{ \left(q(X_i^{[1]})^k = 1/_{m_i} \right), \left(q(X_i^{[2]})^k = 1/_{m_i} \right), \ldots, \left(q(X_i^{[m_i]})^k = 1/_{m_i} \right) \right\} \quad \text{(C.1)}$$

i.e. assign uniform probabilities to the strategies of agent i. This is because, at the beginning, least information is available (largest uncertainty and highest entropy) about which strategy is favorable for the minimization of the collection of system objectives $\sum_{r=1}^{m_i} G(\mathbf{Y}_i^{[r]})$.

© Springer International Publishing Switzerland 2015
A.J. Kulkarni et al., *Probability Collectives*, Intelligent Systems Reference Library 86, DOI 10.1007/978-3-319-16000-9

1.4 Compute the gradient of the Homotopy function in Eq. (2.10) as follows:

$$\mathbf{C}^k = \left[\frac{\partial J_i(q(\mathbf{X}_i), T)^k}{\partial q\left(X_i^{[1]}\right)^k} \quad \frac{\partial J_i(q(\mathbf{X}_i), T)^k}{\partial q\left(X_i^{[2]}\right)^k} \quad \cdots \quad \frac{\partial J_i(q(\mathbf{X}_i), T)^k}{\partial q\left(X_i^{[m_i]}\right)^k} \right]$$
$$= \left[G\left(\mathbf{Y}_i^{[1]}\right) \cdot \prod q\left(X_{(i)}^{[?]}\right) + \frac{T}{\ln(2)} \left[1 + \ln\left(q\left(X_i^{[1]}\right)^k\right)\right] \right.$$
$$G\left(\mathbf{Y}_i^{[1]}\right) \cdot \prod q\left(X_{(i)}^{[?]}\right) + \frac{T}{\ln(2)} \left[1 + \ln\left(q\left(X_i^{[2]}\right)^k\right)\right]$$
$$\left. \ldots \; G\left(\mathbf{Y}_i^{[m_i]}\right) \cdot \prod q\left(X_{(i)}^{[?]}\right) + \frac{T}{\ln(2)} \left[1 + \ln\left(q\left(X_i^{[m_i]}\right)^k\right)\right] \right] \tag{C.2}$$

2. Compute the search direction as $\mathbf{d}_i^k = -\mathbf{C}_i^k.\left(\mathbf{H}_i^k\right)^{-1}$.
3. Compute $\mathrm{J}_i\left(\left(q(\mathbf{X}_i) + \alpha_{step}.\mathbf{d}_i^k\right), T\right)$
4. Update the probability vector $q(\mathbf{X}_i)^{k+1} = q(\mathbf{X}_i)^k + \alpha_{step}.\mathbf{d}_i^k$
5. Update the Hessian $\mathbf{H}_i^{k+1} = \mathbf{H}_i^k + \mathbf{D}_i^k + \mathbf{E}_i^k$

where $\mathbf{D}_i^k = \frac{\mathbf{y}_i^k.\left(\mathbf{y}_i^k\right)^T}{\mathbf{y}_i^k.\mathbf{s}_i^k}$, $\mathbf{E}_i^k = \frac{\mathbf{C}_i^k.\left(\mathbf{C}_i^k\right)^T}{\mathbf{C}_i^k.\mathbf{d}_i^k}$, $\mathbf{s}_i^k = \alpha_{step}.\mathbf{d}_i^k$, $\mathbf{y}_i^k = \mathbf{C}_i^{k+1} - \mathbf{C}_i^k$ and

$$\mathbf{C}^{k+1} = \left[\frac{\partial J_i(q(\mathbf{X}_i), T)^{k+1}}{\partial q\left(X_i^{[1]}\right)^{k+1}} \quad \frac{\partial J_i(q(\mathbf{X}_i), T)^{k+1}}{\partial q\left(X_i^{[2]}\right)^{k+1}} \quad \cdots \quad \frac{\partial J_i(q(\mathbf{X}_i), T)^{k+1}}{\partial q\left(X_i^{[m_i]}\right)^{k+1}} \right]$$
$$= \left[G\left(\mathbf{Y}_i^{[1]}\right) \cdot \prod q\left(X_{(i)}^{[?]}\right) + \frac{T}{\ln(2)} \left[1 + \ln\left(q\left(X_i^{[1]}\right)^{k+1}\right)\right] \right.$$
$$\ldots \; G\left(\mathbf{Y}_i^{[1]}\right) \cdot \prod q\left(X_{(i)}^{[?]}\right) + \frac{T}{\ln(2)} \left[1 + \ln\left(q\left(X_i^{[2]}\right)^{k+1}\right)\right]$$
$$\left. \ldots \; G\left(\mathbf{Y}_i^{[m_i]}\right) \cdot \prod q\left(X_{(i)}^{[?]}\right) + \frac{T}{\ln(2)} \left[1 + \ln\left(q\left(X_i^{[m_i]}\right)^{k+1}\right)\right] \right]$$

6. Accept the current probability distribution $q(\mathbf{X}_i)^k$, if $k \geq v$ or the condition $q(\mathbf{X}_i)^k - q(\mathbf{X}_i)^{k-1} \leq \varepsilon_2$ is true for successive considerable number of iterations, then stop, else update $k = k + 1$ and go to C.2.

As an example, the Homotopy function minimization when solving the Sensor Network Coverage Problem (SNCP) is presented in Fig. C.1. The initial uniform probability distribution, the intermediate iteration distributions, and the converged probability distribution clearly distinguishing the contribution of every strategy from one of the runs solving the SNCP discussed in Sect. 7.2.2 are illustrated in Fig. C.2.

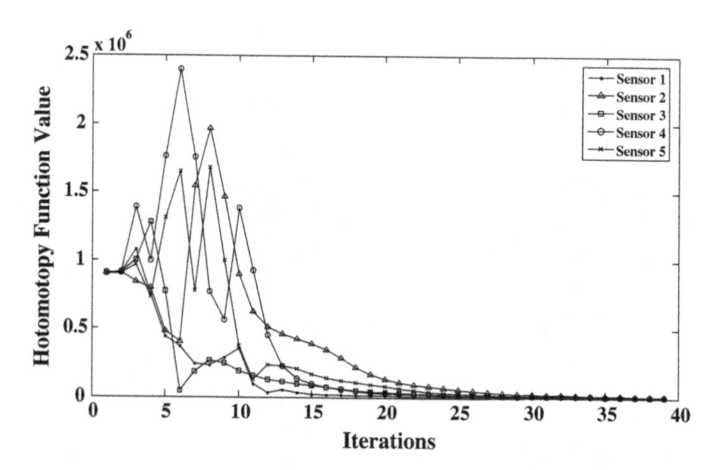

Fig. C.1 Convergence of the homotopy functions for the SNCP using BFGS method

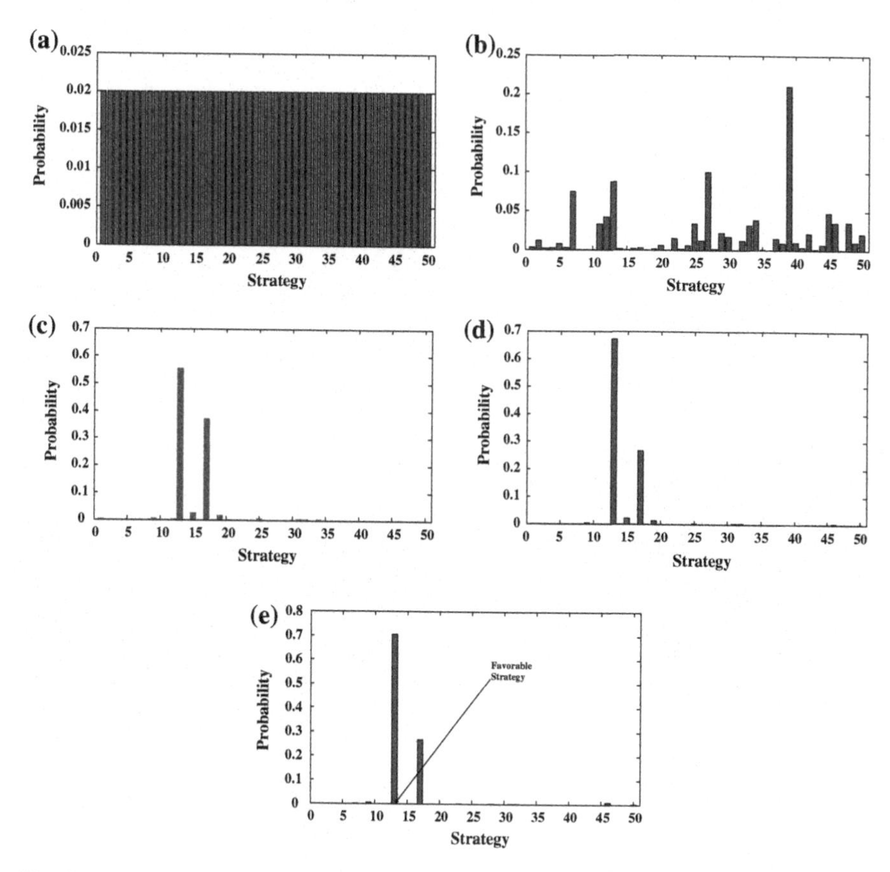

Fig. C.2 Solution history of probability distribution using BFGS method

Appendix D
Individual Sensor Coverage Calculation

There are four possible cases of partial coverage of square FoI in which the sensing area of the sensor i crosses one or more boundaries of the FoI. The four cases associated with it are discussed below with corresponding illustrations presented from Figs. D.1, D.2, D.3 and D.4.

(a) Case 1: The sensor near a corner of the square FoI with the corner outside the sensing area

The area of Segment 1 $A_{Segment\,1}$ bounded by arc $\widehat{1\,2}$ associated with $\angle\beta_1$ and chord $\overline{1\,2}$, and area of Segment 2 $A_{Segment\,2}$ bounded by arc $\widehat{3\,4}$ associated with $\angle\beta_2$ and chord $\overline{3\,4}$ are calculated as follows:

$$A_{Segment\,1} = \left(r_s^2(\beta_1 - \sin\beta_1)\right)/2 \tag{D.1}$$

where

$$\beta_1 = \cos^{-1}\left[\left(d(i,1)^2 + d(i,2)^2 - d(1,2)^2\right)/\left(2 \times d(i,1)^2\right)\right] \quad \text{and}$$

$$d(i,1) = r_s$$
$$d(i,2) = r_s$$
$$d(1,2) = x_2 - x_1,$$

$$A_{Segment\,2} = \left(r_s^2(\beta_2 - \sin\beta_2)\right)/2 \tag{D.2}$$

where

$$\beta_2 = \cos^{-1}\left[\left(d(i,3)^2 + d(i,4)^2 - d(3,4)^2\right)/\left(2 \times d(i,3)^2\right)\right]$$

© Springer International Publishing Switzerland 2015

A.J. Kulkarni et al., *Probability Collectives*, Intelligent Systems Reference Library 86, DOI 10.1007/978-3-319-16000-9

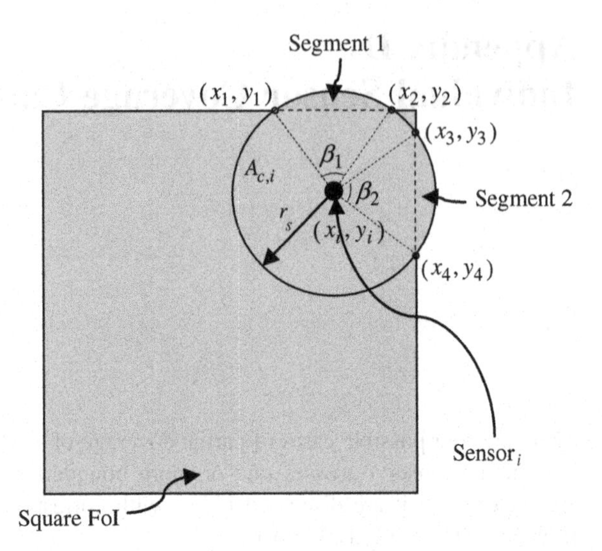

Fig. D.1 Case 1

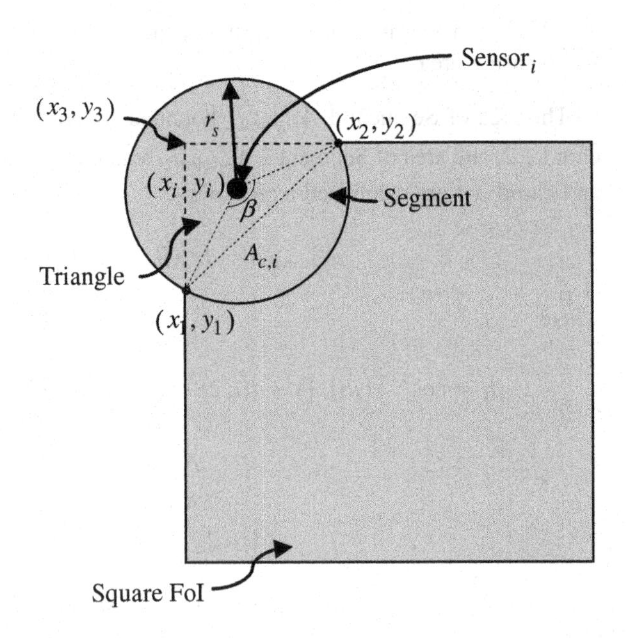

Fig. D.2 Case 2

$$d(i,3) = r_s$$
$$d(i,4) = r_s$$
$$d(3,4) = y_3 - y_4$$

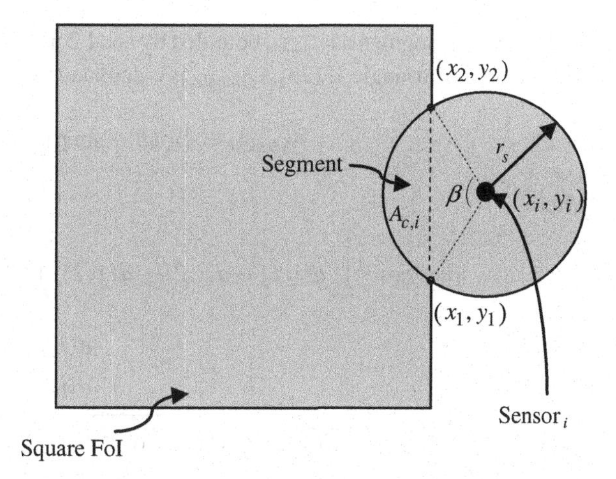

Fig. D.3 Case 3

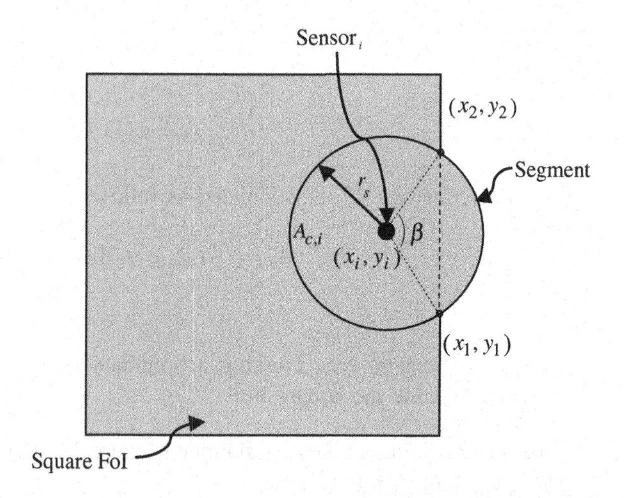

Fig. D.4 Case 4

The FoI coverage $A_{c,i}$ is calculated as follows:

$$A_{c,i} = \pi r_s^2 - \left(A_{Segment\,1} + A_{Segment\,2}\right) \tag{D.3}$$

(b) Case 2: The sensor near a corner of the square FoI with the corner inside the sensing area

The area of Segment $A_{Segment}$ bounded by arc $\widehat{1\,2}$ associated with $\angle\beta$ and chord $\overline{1\,2}$ and the area of triangle $\Delta 123$, $A_{Triangle}$ is calculated as follows:

$$A_{Segment} = \left(r_s^2(\beta - \sin\beta)\right)/2 \tag{D.4}$$

where

$$\beta = \cos^{-1}\left[\left(d(i,1)^2 + d(i,2)^2 - d(1,2)^2\right)/\left(2 \times d(i,1)^2\right)\right] \tag{D.5}$$

$$d(i,1) = r_s$$
$$d(i,2) = r_s$$
$$d(1,2) = \sqrt{(x_1 - x_2)^2 + (y_1 - y_2)^2}$$

$$A_{Triangle} = (d(1,3) \times d(2,3))/2 \tag{D.6}$$

where

$$d(1,3) = y_3 - y_1$$
$$d(2,3) = x_2 - x_3$$

The FoI coverage $A_{c,i}$ is calculated as follows:

$$A_{c,i} = A_{Triangle} + A_{Segment} \tag{D.7}$$

(c) Case 3: Sensing area crossing a boundary of the square FoI at a side with sensor outside the square FoI

The area of Segment $A_{Segment}$ bounded by arc $\widehat{1\,2}$ associated with $\angle\beta$ and chord $\overline{1\,2}$ is calculated as follows:

$$A_{Segment} = \left(r_s^2(\beta - \sin\beta)\right)/2 \tag{D.8}$$

where

$$\beta = \cos^{-1}\left[\left(d(i,1)^2 + d(i,2)^2 - d(1,2)^2\right)/\left(2 \times d(i,1)^2\right)\right] \tag{D.9}$$

$$\begin{aligned} d(i,1) &= r_s \\ d(i,2) &= r_s \\ d(1,2) &= y_2 - y_1 \end{aligned} \tag{D.10}$$

The FoI coverage $A_{c,i}$ is calculated as follows:

$$A_{c,i} = A_{Segment} \tag{D.11}$$

(d) Case 4: sensing area crossing the boundary of the square FoI at a side with sensor inside the square FoI

The area of Segment $A_{Segment}$ bounded by arc $\widehat{1\,2}$ associated with $\angle\beta$ and chord $\overline{1\,2}$ is calculated as follows:

$$A_{Segment} = \left(r_s^2(\beta - \sin\beta)\right)/2 \tag{D.12}$$

where

$$\beta = \cos^{-1}\left[\left(d(i,1)^2 + d(i,2)^2 - d(1,2)^2\right)/\left(2 \times d(i,1)^2\right)\right] \tag{D.13}$$

$$\begin{aligned} d(i,1) &= r_s \\ d(i,2) &= r_s \\ d(1,2) &= y_2 - y_1 \end{aligned} \tag{D.14}$$

The FoI coverage $A_{c,i}$ is calculated as follows:

$$A_{c,i} = \pi r_s^2 - A_{Segment} \tag{D.15}$$